STRATEGIES for CREATIVE PROBLEM SOLVING

Third Edition

LIVERPOOL JMU LIBRARY

3 1111 01433 9475

STRATEGIES for CREATIVE PROBLEM SOLVING

Third Edition

H. Scott Fogler
Steven E. LeBlanc

with

Benjamin R. Rizzo

PRENTICE
HALL

Upper Saddle River, NJ • Boston • Indianapolis • San Francisco
New York • Toronto • Montreal • London • Munich • Paris • Madrid
Capetown • Sydney • Tokyo • Singapore • Mexico City

Many of the designations used by manufacturers and sellers to distinguish their products are claimed as trademarks. Where those designations appear in this book, and the publisher was aware of a trademark claim, the designations have been printed with initial capital letters or in all capitals.

The authors and publisher have taken care in the preparation of this book, but make no expressed or implied warranty of any kind and assume no responsibility for errors or omissions. No liability is assumed for incidental or consequential damages in connection with or arising out of the use of the information or programs contained herein.

The publisher offers excellent discounts on this book when ordered in quantity for bulk purchases or special sales, which may include electronic versions and/or custom covers and content particular to your business, training goals, marketing focus, and branding interests. For more information, please contact:

U.S. Corporate and Government Sales
(800) 382-3419
corpsales@pearsontechgroup.com

For sales outside the United States, please contact:

International Sales
international@pearson.com

Visit us on the Web: informit.com/ph

Library of Congress Cataloging-in-Publication Data

Fogler, H. Scott.

    Strategies for creative problem solving / H. Scott Fogler, Steven E. LeBlanc ; with Benjamin Rizzo.—Third edition.

      pages   cm

    Includes bibliographical references and index.

    ISBN 978-0-13-309166-3 (pbk. : alk. paper)

    1. Problem solving. 2. Creative thinking. I. LeBlanc, Steven E. II. Title.

    BF449.F7 2014

    153.4'3—dc23

                                  2013015512

Copyright © 2014 Pearson Education, Inc.

All rights reserved. Printed in the United States of America. This publication is protected by copyright, and permission must be obtained from the publisher prior to any prohibited reproduction, storage in a retrieval system, or transmission in any form or by any means, electronic, mechanical, photocopying, recording, or likewise. To obtain permission to use material from this work, please submit a written request to Pearson Education, Inc., Permissions Department, One Lake Street, Upper Saddle River, New Jersey 07458, or you may fax your request to (201) 236-3290.

ISBN-13: 978-0-13-309166-3
ISBN-10:   0-13-309166-X

Text printed in the United States on recycled paper at Edwards Brothers Malloy in Ann Arbor, Michigan.

First printing, September 2013

Editor-in-Chief
Mark L. Taub

Executive Editor
Bernard Goodwin

Editorial Assistant
Michelle Housley

Managing Editor
John Fuller

Full-Service Production Manager
Julie B. Nahil

Copy Editor
Christine Dahlin

Proofreader
Linda Begley

Indexer
Ted Laux

Compositor
LaurelTech

Photo Research
Mike Lackey

Permissions Editor
Kathleen Karcher

Dedicated to the memory of Donald R. Woods,
innovative teacher, engineer, scholar, problem solver,
colleague, and friend.

CONTENTS

ADDITIONAL MATERIAL ON THE WEB SITE (WWW.UMICH.EDU/~SCPS):

Power Point Slides of Lectures

Summary Notes

Interactive Computer Games

Material from the Second Edition Not Included in the Third Edition Printed Book

Examples and Exercises That Complement the Examples in the Printed Book

Appendix 1: McMaster's Five-Point Strategy

Appendix 2: Plotting Data

PREFACE

The purpose of this book is to help problem solvers improve their "street smarts." We know that every individual possesses creative skills of one type or another and that these skills can be sharpened if they are exercised regularly. This book provides a framework to hone and polish these creative problem-solving skills.

Strategies for Creative Problem Solving is intended for students, new engineers, practitioners, or anyone else who wants to increase his or her problem-solving skills. After studying this book, you will be able to think critically about an ill-defined problem you encounter, identify the real problem, generate and implement solutions, and then evaluate your solutions. You will develop the skills needed to achieve these goals by examining the components of a problem-solving algorithm and studying a series of graduated exercises intended to familiarize, reinforce, challenge, and stretch your creativity in the problem-solving process.

To cut through the maze of obstacles blocking the pathway to the solution to a problem, we need skills analogous to a pair of scissors with two special blades.

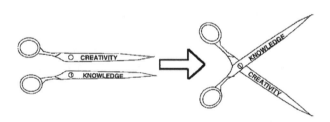

One of the shears is made of the **knowledge** necessary to understand the problem and to develop technically feasible solutions. Of course, no cutting can be done—and no problems of invention can be solved—with just one shear. The other shear contains **creativity** that can generate new and innovative ideas. Likewise, creativity alone will not necessarily generate solutions that are technically feasible— and no cutting can be done with just this single shear, either. Instead, the combination of creativity with a strong technical foundation allows us to cut through the problem to obtain original solutions.

With the aid of a major grant from the National Science Foundation, we researched the wide variety of problem-solving techniques used in industry. Teams of students and faculty visited a number of companies (see the list the Acknowledgments section) to study their problem-solving strategies. We also carried out an extensive survey of new employees, experienced engineers, and managers in industry to collect information on the problem-solving process. As a result of our research, which we share in this text, we know you can become a better problem solver.

A number of engineers and managers provided examples of industrial problems that were incorrectly defined. These examples of ill-defined problems vividly illustrate the need to define the *real* problem as opposed to the *perceived* problem. We believe that if a problem-solving heuristic had been applied to some of these problems in the first place, then the real problem would have been uncovered more rapidly.

A problem-solving heuristic is a systematic approach to problem solving that helps guide us through the solution process and generate alternative solution pathways. The heuristic in this book is quite robust and is applicable to many types of problems. The problem-solving techniques presented in this book provide an organized, logical approach to generating more creative solutions. This book is designed to guide you step by step through the problem-solving process.

Chapter Objectives

The chapter-by-chapter objectives for this new edition of the text are shown below. They provide a logical pathway to help improve and exercise your problem-solving skills.

- **Chapter 1, Problem-Solving Strategies: Why Bother?** This chapter documents several examples where people defined and proceeded to solve the *perceived problem* instead of defining and solving the *real problem*. This chapter also introduces a problem-solving heuristic that can help define the real problem.
- **Chapter 2, The Characteristics, Attitudes, and Environment Necessary for Effective Problem Solving.** This chapter takes an introspective look at the characteristics, habits, and actions that effective problem solvers use, and it helps readers develop the skills necessary to be effective and productive members of a team that is working together to solve problems.
- **Chapter 3, Skills Necessary for Effective Problem Solving.** This chapter focuses on critical thinking, structured critical reasoning, Socratic questions, and critical thinking actions. These skills are essential in defining the problem and in effective problem solving.
- **Chapter 4, First Steps.** This chapter describes different methods to gather information to define and solve the real problem. These techniques include collecting and analyzing information, talking with people who are familiar with the problem, viewing the problem firsthand, and confirming all findings.
- **Chapter 5, Problem Definition.** This chapter provides four techniques you can use to help ensure that you have defined the real problem instead of the perceived problem: critical thinking, the Duncker diagram, the statement–restatement technique, and Kepner–Tregoe (K.T.) problem analysis.

- **Chapter 6, Breaking Down the Barriers to Generating Ideas.** This chapter provides techniques that will help you break down barriers and preconceived notions that get in the way when you are trying to generate solutions to the problem. It also suggests ways to try to enhance your risk taking, because most truly innovative solutions require some risk.
- **Chapter 7, Generating Solutions.** This chapter provides a number of techniques to help you generate solutions to the correctly defined problem, including brainstorming, vertical and lateral thinking, cross-fertilization, futuring, analogy, and TRIZ.
- **Chapter 8, Deciding the Course of Action.** This chapter provides algorithms that will help you decide which problem you should work on first, which solution you should choose, how to identify potential problems, and how you can prevent problems from occurring in the future.
- **Chapter 9, Implementing the Solution.** This chapter describes the steps necessary to implement the decisions you made using the techniques described in Chapter 8.
- **Chapter 10, Evaluation.** This chapter shows you how to evaluate the solution you have implemented, making sure it completely solves the problem, it is ethical, and it is safe for both people and the environment.
- **Chapter 11, Troubleshooting.** This chapter provides an algorithm and a worksheet that you can use for troubleshooting problems of malfunctioning industrial equipment and systems. This algorithm also applies to nontechnical problems.
- **Chapter 12, Putting It All Together.** This chapter shows how all of the problem-solving techniques presented in Chapters 1 through 11 can be applied to two real-life case studies. The first case study focuses on a project carried out by Engineers Without Borders (EWB) and was provided by Dr. Marina Miletic. The second case study on finding and solving workplace problems at Bakery Café is from a term project in a senior-level problem-solving course at the University of Michigan.

What's New in the Third Edition?

Chapter 3 introduces the concept of *structured critical reasoning* (SCR) and how to apply it to different situations to see if the conclusions are supported by evidence and do not contain any of the 11 fallacies in logic. In this chapter, a greater emphasis is placed on how to ask critical-thinking questions and carry out critical-thinking actions in order to define and solve the *real problem* instead of the *perceived problem*.

Other new material discusses how to manage complex change. Many of the explanations of the components of the problem-solving heuristic have been expanded and clarified in most of the chapters.

Many of the examples used to illustrate the different components of our problem-solving heuristic are new. New examples include "A Picture Is Worth a Thousand Words," "SCR: Truancy in U.K. Schools," "SCR: Draft Once Again" (on the Web site), "SCR: A Public Health Hazard—Eggs," "Concerns about a New Energy Drink," "Blind to the Cause," "Everyone's Private Driver," "Fermi Example: Piano Tuners in Chicago," "The Anti-Crime Commissioner," "Viewing the Problem Firsthand," "Bitrex," "Dead Rats Decaying in the Water," "Overcoming Fear of Failure: Nokia's Bad Call on Smartphones," "Futuring in Action: Highway Congestion No More," "Virgin Atlantic Airlines: Entertainment on the Go," "Cross-Fertilization in Practice: Eradicating Malaria," "Increasing the Size of the Boeing 737," "Empty Soapboxes," "K.T. Situation Appraisal: Flooding in Thailand," "Decision Analysis: Popara Enterprises Picks a New Market," "K.T. Potential Problem Analysis: Jet-Lagged John," "25 Years or Less?," "Changing Buses," "Adapting during the Problem-Solving Process: Better TV," "Evaluation of Bird Droppings and Car Paint," "Troubleshooting: The Morning Newspaper (Based on a True Story)," and "Troubleshooting: Gold Production."

The Web Site

The Web site (www.umich.edu/~scps) provides a number of learning resources:

- Summary Notes
- Microsoft PowerPoint Slides of Lecture Notes
- Interactive Computer Modules
- Professional Reference Shelf

The Summary Notes contain highlights from lectures on each chapter used in a senior-level problem-solving course, whose syllabus can be accessed from the

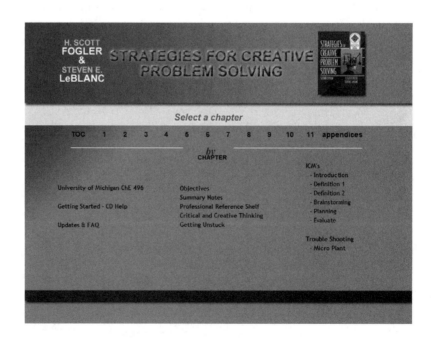

home page. The Web material contains such things as FAQs, Chapter Objectives, Critical and Creative Thinking, and Thoughts on Problem Solving. The Professional Reference Shelf offers material from the first and second editions along with additional examples and material for each chapter.

Mike Kravchenko initially developed the Web site by incorporating all of the Summary Notes and the material found in the Professional Reference Shelf. Dr. Nihat Gürmen developed the format for the Web site and supervised the development of the interactive computer modules. Nihat's and Mike's creativity and attention to detail to make everything consistent were amazing. For this edition, Maria Quigley worked on the Web site and Arthur Shih brought the site to its final polished form.

Slides from the Summary Notes can be used for classroom presentations.

Seven interactive computer modules are available to supplement and reinforce material in the text.

- **Introduction** (Chapters 1 and 2): An interactive puzzle/game that stresses the foundations of problem solving to help you learn the concepts.
- **Definition 1** (Chapter 5): This module reviews and offers practice with problem definition techniques.
- **Definition 2** (Chapter 5): You can sharpen your problem definition skills by working on one of three scenarios involving vague problem statements.
- **Brainstorming** (Chapter 7): This module leads you through a variety of brainstorming and blockbusting techniques.
- **Planning** (Chapter 9): This module helps you practice implementation of a problem solution by using a student bridge-building competition scenario.
- **Evaluate** (Chapter 10): Playing the part of an employee in a paper mill, you are asked to evaluate a proposed plant expansion.
- **Troubleshooting a Micro Plant** (Chapter 11): Actual equipment malfunctions from chemical plants were collected and incorporated into this model of a plant manufacturing ethyl benzene. You must troubleshoot the plant to identify the two faults that are randomly generated each time you sign on. Additional information on process equipment is available on the Web site at http://encyclopedia.che.engin.umich.edu.

ACKNOWLEDGMENTS

A number of colleagues and students contributed greatly to this edition and we are most grateful. Nikola Popara and Alexander Olsson contributed to and polished some of the examples. Laura Swierzbin proofread the final draft of the manuscript; Maria Quigley provided the Dow Corning troubleshooting problem. Arthur Shih put significant effort into developing the final version of the Web site. His always cheerful disposition in carrying out this development was appreciated. Wattana (Zack) Chaisoontornyotin, Thammaporn (Pearl) Somkhan, and Thanawat (Tum) Thanthong contributed to the "Flooding in Thailand" example. The material on structured critical reasoning was developed with the guidance of Professor Marco Angelini while one of the authors (HSF) was on sabbatical at University College London (UCL). Our lunch meetings and discussions contributed greatly to the third edition of *Strategies for Creative Problem Solving* (*SCPS*) in many ways. Professor Angelini (now at University of Technology, Sydney) provided help and encouragement that are greatly appreciated. Others at UCL whom HSF would like to thank are Professors David Bogle and Asterios Gavriilidis, as well as the UCL chemical engineering office staff—Pat Markey, Mae Oroszlany, and Agata Blaszczyk—who facilitated his stay at UCL in London. The material on "77 Cards: Design Heuristics" was developed by a University of Michigan colleague, Dr. Shanna Daly. We would like to thank Don Woods and Tom Marlin of McMaster University for sharing their troubleshooting example with us and for Don's continual encouragement and pioneering work on teaching troubleshooting. David Zinn provided many of the drawings and in a number of cases suggested the drawing from the written description of the scenario or case study.

   Colleagues who read the final draft of this edition and gave valuable comments include Professor John Falconer at the University of Colorado, Dr. Susan Montgomery at the University of Michigan, and Dr. Gavin Towler, CTO of UOP. Other colleagues who provided suggestions are Hank Kohlbrand, Dow Chemical (retired); June Wispelwey, Executive Director, American Institute of Chemical Engineers; and Sid Sapake, General Mills (retired).

   The authors are greatly appreciative of the detailed comments given on each chapter by the Engineering 405 course during the 2012 fall term. They are Alex Benfield, Emily Caoagas, Victoria Choe, Daniel Friedman, Paul Hillman, Ching-Han Hsieh, Sahil Jain, Joel Kaatz, Carissa Le, Matt Mackey, Arjun Mahajan, Christopher McMullen, Jordan Morgan, Zachary Mutual, Michael O'Connor, Jacob Obron, Jeffrey Petsch, Chloe Prince, Otis Putt, Radhisha Rughani, Travis Scadron, Anusha Sthanunathan, and Laura Swierzbin. Laura Swierzbin, Apeksha Bandi, Mayur Tikmani, Gustav Sandborgh, Emily Caoagas, and Corey Pederson worked on the Instructor's Solutions Manual. Claudio Vilas Boas Favero contributed a number of suggestions and did an excellent job as the graduate student instructor, working with these students to make the class and term projects a success. The syllabus and PowerPoint slides of the lectures for this class can be found on the Web site for this book. We also point the readers to take note of the following cryptogram: However, all intensive laws tend often to have exceptions. Very interesting concepts take orderly, responsible statements. Virtually all laws intrinsically are natural thoughts. General observations become laws under experimentation.

The production staff at Prentice Hall was outstanding, as usual. We also would like to recognize our publisher, Bernard Goodwin; our production editor, Julie Nahil; and our copy editor, Christine Dahlin, for their excellent work.

As in the last two editions of *SCPS*, we are most indebted to Ms. Laura Bracken again for all the energy and effort she put in. Not only did she type and format the entire book, but she also gave many suggestions on a variety of things, corrected inconsistencies, and took a proactive approach on every aspect of the book. Laura, thank you so much.

One person who deserves special recognition for her work in making this book possible is Janet Fogler, our de facto editor. Janet spent endless hours going through every page of the manuscript, rewriting sentences, making sure things followed logically, giving us suggestions on how to make things clearer, and helping in many other ways. She took our original manuscript and raised it to a higher level and, indeed, is an integral part of this work.

We wish to thank the following companies for participating in this project to provide real examples and share their problem-solving methodologies:

Amoco	KMS Fusion
Chevron Specialty Chemicals	Kraft General Foods
Dow Chemical	Mobil
Dow Corning	Monsanto
DuPont	Procter & Gamble
Eli Lilly	Shell
General Mills	3M Upjohn

H. Scott Fogler, Ann Arbor, Michigan
Steven E. LeBlanc, Toledo, Ohio
Benjamin R. Rizzo, Pittsburgh, Pennsylvania
August 2013

Strategies for Creative Problem Solving Web site:
www.umich.edu/~scps

ABOUT THE AUTHORS

 **H. Scott Fogler** is the Ame and Catherine Vennema Professor of Chemical Engineering, an Arthur F. Thurnau Professor at the University of Michigan, and the 2009 President of the American Institute of Chemical Engineers (AIChE). Fogler has chaired ASEE's Chemical Engineering Division. He also has earned the Warren K. Lewis Award from AIChE for contributions to chemical engineering education and the 2010 Malcolm E. Pruitt Award from the council for Chemical Research. He is the author of the best-selling textbook *Elements of Chemical Reaction Engineering,* now in its fourth edition (Prentice Hall, 2006), and the textbook *Essentials of Chemical Reaction Engineering* (Prentice Hall, 2010).

 **Steven E. LeBlanc** is Executive Associate Dean for Academic Affairs and Professor of Chemical Engineering at the University of Toledo. He previously served as the chair of the Chemical and Environmental Engineering Department for ten years. LeBlanc has served as the chairman of the ASEE Chemical Engineering Education Division and as a co-chair of the 2007 ASEE Chemical Engineering Summer School for Faculty.

 **Benjamin R. Rizzo** is a production engineer for Shell Oil, working on Unconventional Reservoir Optimization in Pittsburgh, Pennsylvania. He is a graduate of the University of Michigan, where he received his B.S.E. in Chemical Engineering with highest honors. At Michigan, Rizzo was the Paul S. Bigby Class of 1931E Scholar and was awarded a Landis Prize for Technical Communication. He was also a member of a whimsical men's octet singing group, the Friars; a member of the Men's Glee Club; and the recipient of the A. Elaine Bychinsky Award for contribution to the campus music community.

1 PROBLEM-SOLVING STRATEGIES: WHY BOTHER?

The ability to successfully solve a given problem is an essential skill in life. Everyone is called upon to solve problems every day—from such mundane decisions as what to wear or where to go for lunch, to the much more difficult problems that are encountered in school or on the job, the outcomes of which often have significant impacts on careers. Most real-world problems have several possible solutions. More complex problems have a large number of alternative solutions, and a variety of drastically different approaches are often plausible for implementation. The goal is to find, select, and implement the best solution for the correctly defined problem. All of us will be better able to achieve this goal if we exercise our problem-solving skills frequently to make them sharper. By understanding and practicing the techniques discussed in this book, you will develop problem-solving "street smarts" and become a much more efficient problem solver.

WHAT'S THE REAL PROBLEM?

With the aid of a grant from the National Science Foundation, the authors of this book carried out an extensive study of problem solving within corporations. Teams of students and faculty from the University of Michigan visited many U.S. corporations to learn how employees and managers approached problem solving, to collect their problem-solving techniques, and to learn about the difficulties that occurred in solving real-world problems. When we summarized the results, we found that the greatest challenge was making sure that the **real problem** was defined instead of the **perceived problem**. The difference between a perceived problem and a real problem is illustrated in the following humorous example.

The Case of the Hungry Grizzly Bear[1]
or
An Exercise in Defining the "Real Problem"

A student and his professor are backpacking in Alaska when a grizzly bear starts to chase them from a distance. Both start running, but it's clear that eventually the bear will catch up with them. The student takes off his backpack, gets his running shoes out, and starts putting them on. His professor says, "You can't outrun the bear, even in running shoes!" The student replies, "I don't need to outrun the bear; I only need to outrun you!"

Of course, the student quickly realized that the perceived problem was to outrun the bear while the real problem was to not get caught by the bear and that the bear would be satisfied with one "catch," so he only had to outrun his professor. This example illustrates a very important issue: **problem definition**.

Problem definition is a common but difficult task because true problems are often disguised in a variety of ways. It takes a skillful individual to analyze a situation and extract the real problem from a sea of information. Ill-defined or poorly posed problems can lead you down the wrong path to a series of impossible or spurious solutions. Correctly defining the real problem is critical to finding a workable solution.

Sometimes we can be "tricked" into treating the symptoms instead of solving the root problem. Treating symptoms (e.g., putting a bucket under a leaking roof) can give us the satisfaction of a quick fix (e.g., no puddle on the floor), but finding and solving the real problem (i.e., the cause of the leak) are important to minimize lost time, money, and effort. Implementing real solutions to real problems requires discipline, and sometimes stubbornness, to avoid being pressured into accepting a less desirable quick-fix solution because of time constraints.

The next few pages present a series of real-life examples from case histories that show how easy it is to fall into the trap of defining and solving the wrong problem. In these examples and the discussions that follow, the perceived problem refers to a problem that is thought to be correctly defined but that actually is not. These examples provide evidence of how millions of dollars and thousands of hours can be wasted as a consequence of poor problem definition and solution.

Impatient Guests

Shortly after the upper floors of a high-rise hotel had been renovated to increase the hotel's room capacity, guests complained that the elevators were too slow. The building manager assembled his assistants. His instruction to solve the **perceived problem** was "find a way to speed up the elevators."

After calling the elevator company and an independent expert on elevators, it was determined that nothing could be done to speed up the elevators. The manager then issued new directions: "Find a location and design a shaft to install another elevator." An architectural firm was hired to carry out this request.

Ultimately, neither the shaft nor the new elevator was installed because shortly after the firm was hired, the **real problem** was uncovered. The real problem statement was "find a way to keep guests happy with the current elevators." Instead of putting effort into decreasing wait time, effort was placed on taking the guests' minds off their wait. The guests stopped complaining when mirrors were installed on each floor in front of the elevators.[2] Few people can resist taking the time to check or admire their appearance in the mirror.

Better Printing Inks

In 1990, the Bureau of Engraving and Printing (BEP) initiated a program to redesign and improve the quality of paper money being printed in the United States. When the design was completed and the new printing machines were installed, a few trial printings were carried out. Unfortunately, it was found that the ink on the new bills smeared when touched. The following instruction was given to solve the **perceived problem**: "Get better printing inks."

A number of workshops and panels were convened to work on this problem. After a year and a half of hard work by both government officials and college faculty on the perceived problem, research programs at several universities were chosen to try to develop better printing inks.

Just as these programs were to be initiated, the BEP withdrew the funds, stating that the real problem was not with the inks but rather with the printing machines. The new machines were not putting sufficient pressure on the new type of paper to force the ink down into the paper far enough to avoid smearing. Consequently, the funds earmarked for research on inks were diverted to the purchase of new printing machines.

By originally defining the wrong problem, the BEP wasted thousands of hours of effort by government officials and college faculty. The **real problem** statement should have been "Find out why the ink is smearing."

Making Gasoline from Coal[3]

A few years ago a major oil company was developing a process for the Department of Energy to produce liquid petroleum products from coal in order to reduce U.S. dependence on foreign oil. In this process, solid coal particles were ground up, mixed with solvent and hydrogen, then passed through a furnace heater to a reactor that would convert the coal to gasoline (see the figure below). After installation, the process was not operating properly. Excessive amounts of a tar-like carbonaceous material were being deposited on the pipes in the furnace, fouling and in some cases plugging the pipes.

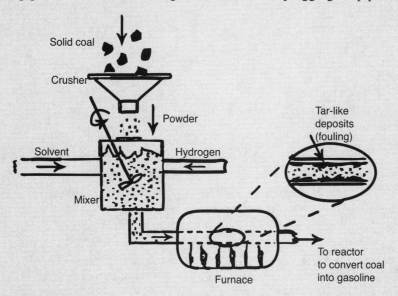

Luiz Rocha/Shutterstock

The instruction given by the manager to his research group to solve the **perceived problem** was to "improve the quality of the solvents used to dissolve the coal and prevent these tar-like deposits." A major research program was initiated. After a year and a half of effort was spent on the process, no one solvent proved to be a better solution to the problem than any other. Perhaps a more general problem statement such as "Determine why the carbon deposits are forming and how they can be eliminated" would have revealed the true problem. The **real problem** was that the velocity of the mixture through the hot pipe in the furnace was so slow that the temperature of the mixed particles and solvent in the pipe became high enough to cause them to react and form a coal-tar-like substance that was then building up on the pipe wall. The problem was solved by using a smaller diameter pipe while maintaining the same total flow rate to increase the velocity through the furnace pipe, so that the particles and solvent had less time to react in the furnace to form the tar-like deposits. In addition, the high velocity caused the coal particles in the fluid to act as scouring agents on the furnace pipe wall. After the furnace pipe was changed, no further problems of this nature were experienced.

A Picture Is Worth a Thousand Words[4]

Cell phone makers, working to increase sales, have steadily increased the quality of phone cameras without substantial increases in cost, and for years, it has been predicted that smartphones would adversely affect digital camera sales. So how did the digital camera industry respond to the problem of this competition? The **perceived problem** was "How could camera makers produce a less expensive camera to compete with cell phone cameras?"

The real problem, however, was not how to produce a cheaper camera but how to produce a higher-quality camera that a consumer would want to buy in addition to a lower-quality phone camera. In January 2012, Japan, the largest producer of digital cameras, shipped three times the number of digital cameras than it did in January 2003, when cell phone cameras were just being introduced. Why? The camera industry has kept a step ahead of smartphones by consistently producing cameras with features not available on cell phones, such as detachable lenses and high optical zooms of 30X and higher. Photography is a very personal activity that preserves important moments in one's life: graduations, weddings, the birth of children, a baby's first steps, etc. Consumers want the best picture possible to capture these important moments, and Nikon, Canon, Olympus, and other camera brands still offer the best technology to do so. The camera industry likely would have lost competing on cost, but it has continued to win by offering superior quality.

CORRECT PROBLEM DEFINITION/WRONG SOLUTION

In this section, we discuss some examples where the real problem has been correctly defined, but the solutions to the problem were woefully inadequate, incorrect, or unnecessary. The individuals who made the decisions in the situations described here were all competent, hardworking professionals. Unfortunately, they overlooked some essential details that might have prevented the accidents and mistakes. Using 20/20 hindsight, consider whether the following situations could have been avoided if an organized problem-solving approach had been applied.

Dam the Torpedoes or Torpedo the Dam?

The Australian government wanted to increase agricultural production by finding ways to grow crops on wastelands where it was currently unable to grow anything. It was decided to cultivate land in the New South Wales region of southeastern Australia, which is very arid. Some wild plants could be seen growing in the soil from time to time, but there was insufficient moisture to grow crops. It was believed that the land could be irrigated and that agricultural food crops could be grown. The Murray River, which flows naturally from the mountains to the sea, passes through the region. The solution chosen by the Australian government to solve the problem of not being able to grow crops in wastelands of New South Wales was this: "Design a system to divert the Murray River inland to irrigate the land and grow agricultural crops."

A series of multimillion-dollar dams were built, and the water was diverted. Unfortunately, when the irrigation was achieved, absolutely no new vegetation grew, and even the vegetation that had previously grown on some of the land died. It was determined that this infertility of the soil occurred because the diverted water dissolved abnormally high concentrations of salts present in the soil, which then entered the plant roots. Little of the vegetation could tolerate the salts at such high concentrations; as a result, even the existing vegetation died. A potential problem analysis (see Chapter 8) might have prevented this costly experiment.[5] Currently, efforts are under way to deal with this salinity problem, with proposed solutions ranging from desalination to the construction of salt ponds.

What's the Disease?

On a lighter note, we end with the following true example of right problem/ wrong solution. At an American Medical Association convention on respiratory diseases a number of years ago, an upper-body X-ray was displayed at the booth of one of the pharmaceutical vendors.

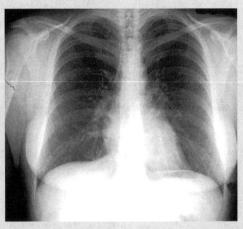

The following instructions were given to the physicians: "Diagnose the ailment from the X-ray, and place your answer in the contest box near the display" (a correct problem statement). The winner of a valuable prize would be drawn from those who had made the correct diagnosis. Because of the focus on the upper torso, virtually every known lung disease treatment was suggested by one physician or another. There was no need to hold a drawing from the correct diagnoses submitted because only one person discovered the true solution: "Set a broken right arm."[6]

Nearly all project design failures, such as those described in the preceding examples, result from faulty judgments rather than faulty calculations. While there is nothing we can do that will guarantee that you will never make mistakes or faulty judgments, we believe that if you use the methods and techniques discussed in this book, you will be less likely to do so.

Pontus Edenberg/Shutterstock

A HEURISTIC FOR SUCCESSFUL PROBLEM SOLVING

In the preceding sections, you saw several examples of ill-defined and incorrectly solved problems. The people who defined these perceived problems were bright and conscientious. So how could they have made these mistakes? And how can we avoid the same pitfalls that ensnared the people in these examples? The goal of this book is to structure the process of defining and solving real problems in a way that will be useful to you in everyday life, both on and off the job. We shall achieve this goal by providing a structure to the problem-solving process called a **heuristic.**

A problem-solving heuristic is a systematic approach that helps guide us through the solution process and generate alternative solution pathways. While a heuristic cannot prevent people from making errors, it provides a uniform, systematic approach to deal with any problem. A heuristic is analogous to a set of directions from one place to another. Your problem is getting from point A to point B, and different heuristics will tell you slightly different routes to travel (for example MapQuest vs. Google Maps). While there is no unique or preferred way to solve these types of problems, we believe the use of a heuristic is an effective technique.

Building Blocks of the Heuristic

The five-step problem-solving heuristic that we will use contains these steps: define, generate, decide, implement, and evaluate. It finds its origins in the McMaster Five-Point Strategy.[7]

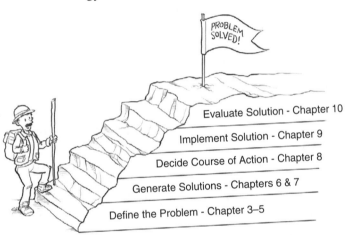

The Five Steps of the Problem-Solving Heuristic

Problem definition, the first step in our problem-solving heuristic, can resurface at any point of the problem-solving process as we encounter dead ends or as the criteria or conditions change. First we gather information, and then we use several specific techniques, such as the Kepner–Tregoe (K.T.) problem analysis technique, to help **define** the problem correctly. In Chapters 3 through 5 we present material and techniques that will help you to define *the real problem* instead of *the perceived problem.*

Once you have defined the real problem, you can proceed to the next step in the heuristic, **generate** solutions to solve the problem. One of the most popular techniques used in industry to generate ideas is brainstorming. In addition to brainstorming, other methods to facilitate idea generation include futuring, analogies, and blockbusting. As part of the idea generation step, it is important to ask questions such as the following:

- What *is* and what *is not* possible?
- Has this problem been solved before?

- Is it worth solving?
- Which resources (e.g., past experience, time, money, personnel) are available to obtain a solution?

In Chapters 6 and 7 we present methods that will help you practice and enhance your creativity and to remove barriers to generating ideas.

The next building block, **decide,** is discussed in Chapter 8 in the context of the Kepner–Tregoe (K.T.) approach.[8] *K.T. situation appraisal* helps you to *decide* which project to work on first. *K.T. problem analysis* helps you to find the root cause of the problem. *K.T. decision analysis* is where you generate a number of solution alternatives, and you then *decide* which alternative to choose. In K.T. potential problem analysis, you plan to ensure the success of your *decision* by identifying everything that might potentially go wrong, the causes of each potential problem, the preventive actions that could be taken, and the steps of last resort.

Having made the decision and planned for its success, you are now ready to **implement** the solution. The first implementation step is to plan the activities you need to perform to solve the problem. Chapter 9 presents a number of techniques to allocate time and resources to help you carry your chosen solution through to its successful completion.

In the **evaluation** phase, you look back and make sure all of the criteria in the problem statements were fulfilled and that none of the constraints were violated. The evaluation phase is carried out both at the end of and during the heuristic, especially when major decisions are made or when branch points occur. Chapter 10 discusses what must be done to evaluate your solution by asking the following questions: Has the problem *really* been solved, and is the solution you chose the best solution? Is the solution innovative, new, and novel, or is it merely an application of existing principles (which, in some cases, may be all that's necessary)? Is the solution ethical, safe, and environmentally responsible?

The heuristic we have just presented, visualized above, will serve as a handy road map for your journey through the problem-solving process. Of course, even when armed with a good road map, travelers may sometimes arrive at the wrong destination or take an excessive amount of time to reach the correct destination. In a similar way, problem solvers may come to the wrong solution or take too much time to obtain a solution. Travelers also have to approach the trip with a positive attitude and draw upon the knowledge of expert travelers who have navigated the road before them. Finally, they need to make sure there is agreement on the route among themselves—conflicts between the travelers can make the trip unpleasant, sometimes with a disastrous outcome. Similarly, problem-solving groups can be much more efficient when they work well together in teams.

In the problem-solving process, one of the first things we want to do is to get in the right frame of mind so we can approach the problem with a positive, enthusiastic attitude. In Chapter 2, we discuss how to develop this mindset when problem solving. Here we present attributes and characteristics of experienced problem solvers, the seven habits of highly effective people, seven actions necessary for a successful career, the need to approach each problem in a positive manner, and some ideas on how to solve problems successfully in teams. After this chapter, we begin discussing the first block of our problem-solving heuristic.

SUMMARY

Why bother with using a problem-solving strategy? This chapter presented a number of factual case histories that illustrate what happens when the real problem isn't defined or when there is no organized approach to problem solving. In the chapters that follow, we present a heuristic and a number of techniques that can greatly enhance your chances of defining and solving the real problem instead of the perceived problem and identifying potential roadblocks during the problem-solving process.

WEB-SITE MATERIAL (WWW.UMICH.EDU/~SCPS)

- **Learning Resources**
  Summary Notes
  Problem-Solving Heuristic
  PowerPoint Slides for Classroom Use
  Getting Unstuck (accessed through the home page)

- **Professional Reference Shelf**

  1. Defining the Real Problem

Bargain Prices

*"What's the
lowest price?"*

Decreasing Profits

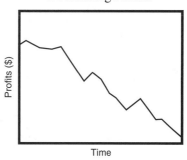

*"Close the plant because the
price of the fertilizer is too low
and we can no longer afford to
operate it."*

2. Correct Problem Definition/Wrong Solution

a) An Unexpected Twist: An Explosion in the Nypro Chemical Plant in Flixborough, England

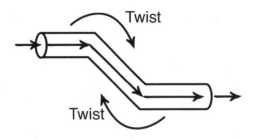

b) The Kansas City Hyatt

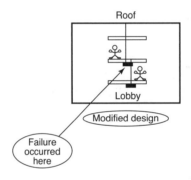

c) Leaking Flow Meter

REFERENCES

1. Prof. John Falconer, University of Colorado, Boulder, CO 80302.
2. Adapted from *Chemtech*, 22, 1, p. 24, 1992.
3. Courtesy of Prof. Antonio Garcia, Arizona State University, Phoenix, AZ 85287-6006.
4. "Camera Defies Smartphone Onslaught, Stays in Frame," *The Huffington Post*, October 24, 2012, and "Still in the Frame, the Camera Defies Smartphone Onslaught," Singapore, April 25, 2012, by Jeremy Wagstaff, Asia chief technology correspondent. Sources: www.huffingtonpost.com/2012/04/25/camerasmartphone_n_1451085.html and www.reuters.com/article/2012/04/25/uk-tech-cameras-id USLNE83O01420120425.
5. www.environment.sa.gov.au/about_us/reports/nrm_annual_reports.
6. Prof. Brymer Williams, University of Michigan, Ann Arbor, MI 48109.
7. Woods, D. R., "The McMaster Five-Point Strategy," notes and personal communication, 1982.
8. Kepner, C. H., and B. B. Tregoe, *The New Rational Manager*, Princeton Research Press, Princeton, NJ, 1981.

EXERCISES

1.1. What were the three most important things you learned in this chapter?

1.2. Think of a recurring problem or situation in your life that you would like to resolve or manage better. Try viewing it as an ill-defined problem, and try to redefine this issue. Do new possibilities come to mind?

1.3. Is there a common thread running through the ill-defined problems on pages 3 through 6? If so, what is it?

1.4. Start to keep a journal or record as we progress through this text on problem solving. Write down things of interest as well as questions that you think will help your problem-solving capabilities. Write down what doesn't work for you as well as what does. List the types of problems you would like to become more skilled at solving.

1.5. Collect two or more ill-defined problems similar to the case histories described in this chapter.

1.6. What are the similarities and differences among the London Underground, Nypro Chemical Plant, Murray River Dam, and Hyatt Regency disasters (discussed in the Professional Reference Shelf on the Web site)?

1.7. Write a paragraph discussing problems you might have with the different building blocks as you use the heuristic discussed in this chapter.

1.8. "Where's the Oil?": Water flooding is a technique commonly used in oil recovery in which water is injected into a well, displacing the oil and pushing it out of another nearby well. In many cases, expensive chemicals are injected along with the water to facilitate pushing out the oil.

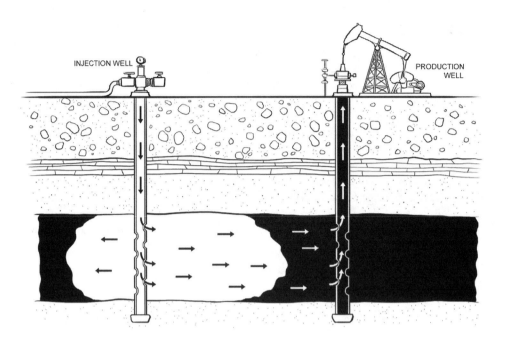

astudio/Shutterstock

A major oil company was having problems with a Canadian light-oil reservoir where the recovery was turning out to be much lower than expected. The following instruction was given to solve the perceived problem: "Find ways to improve the oil recovery rate."

A. Make a list of at least five other "instructions" that could have been given.

B. Make a list of five questions you would have asked.

Problem courtesy of Dr. Mark Hoefner, Mobil Research and Development Corporation, Dallas, TX 75387.

1.9. "Dam the Torpedoes or Torpedo the Dam?": Write four or five sentences listing questions you would have asked before the building of the dams project was initiated.

1.10. Write four or five questions you would have asked before the construction of the new branch of the Underground Jubilee Trains (found on the Web site).

1.11. "Better Printing Inks": Make a list of five questions you would have asked before sending out the request for proposals to universities and industries for a new printing ink (page 4)?

1.12. Visit the problem-solving Web site at www.umich.edu/~scps and write a paragraph describing all the resources on the site and how you could use them.

1.13. In the "Making Gasoline from Coal" example (page 5), at what point or period of time would you have stopped testing new solvents and why would you have stopped?

FURTHER READING

Copulsky, William. "Stories from the Front," *Chemtech,* 22, p. 154, 1992. More anecdotal cases of histories of ill-defined situations and solutions.

2 THE CHARACTERISTICS, ATTITUDES, AND ENVIRONMENT NECESSARY FOR EFFECTIVE PROBLEM SOLVING

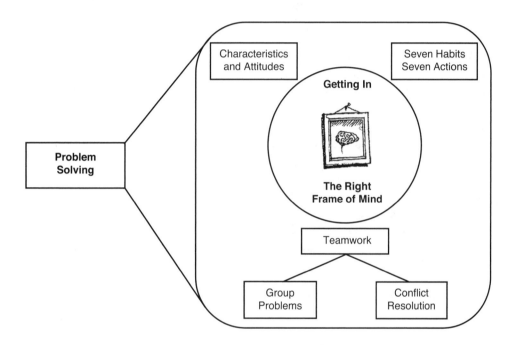

In this chapter we present ways to improve your problem-solving skills both individually and as a team member. We first discuss ways to improve your individual skills—namely, by practicing the habits of effective problem solvers, adopting a positive attitude, and embracing advice from industry on how to be successful in a creative environment. Next we encourage you to take notes, have a vision of the "big picture," and look for paradigm shifts. We close the chapter with a discussion of problem solving in teams and an exploration of ways that team members can work together effectively to solve problems.

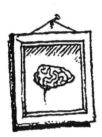

GETTING IN THE RIGHT FRAME OF MIND

Characteristics of Effective Problem Solvers

Extensive research has been carried out in an effort to determine the differences between effective problem solvers and ineffective problem solvers.[1,2] The most important factors that appear to distinguish effective from ineffective problem solvers are the *attitudes* with which they approach the problem, their *aggressiveness* in the problem-solving process, their concern for *accuracy*, and the *solution procedures* they use. For example, effective problem solvers believe that problems can be solved through the use of heuristics and careful persistent analysis. By contrast, ineffective problem solvers think, "You either know it or you don't." Effective problem solvers become very active in the problem-solving process; they draw figures, make sketches, and ask questions of themselves and others. Ineffective problem solvers don't seem to understand the level of personal effort needed to solve the problem. Effective problem solvers take great care to understand all the facts and relationships accurately. Ineffective problem solvers make judgments without checking for accuracy.

Characteristics that students have found particularly useful in solving single-answer, closed-ended problems given as homework include (1) re-describing/rephrasing the problem statement and (2) starting at a point they can first understand even though it may not be what, at first glance, appears the most logical point.

By approaching a situation using the characteristic attitudes and actions of an effective problem solver, you will be well on your way to finding the real problem and generating an outstanding solution (see the Web site for more about characteristics).

*If you think
you can—
you will.*

*If you think
you can't—
you won't.*

Attitudes for Effective Problem Solving

Effective problem solvers develop mindsets and habits that aid them in dealing with difficult problems. Stephen Covey's research on highly effective people revealed that these successful people adopt a similar set of habits.[3] We encourage you to practice the seven habits shown in the following table.

The Seven Habits of Highly Effective People

Habit 1: Be Proactive. 	• Take the initiative and make things happen. • Aggressively seek new ideas and innovations. • Don't let a negative environment affect your behavior and decisions. • Work on things that you can do something about. • If you make a mistake, acknowledge it, learn from it, and move on.

Habit 2: Begin with the End in Mind. 	• Know where you are going, and make sure all the steps you take are in the right direction. • First determine the right things to accomplish and then identify how to best accomplish them. • Write a personal mission statement describing where you want to go, what you want to do, and how you will accomplish these things.
Habit 3: Put First Things First. 	• List your top priorities each day for the upcoming week and schedule time to work on them. • Continually review and prioritize your goals. Say "no" to unimportant tasks. • Focus on the important tasks, the ones that will have the greatest impact if they are carefully thought out and planned.
Habit 4: Think Win/Win. 	• Win/win is the frame of mind that seeks mutual benefits for all people involved in solutions and agreements. • Identify the key issues and results that would constitute a fully acceptable solution to all. • Make everyone who is involved in the decision feel good about the decision and committed to a plan of action.
Habit 5: Seek First to Understand, Then to Be Understood. 	• Learn as much as you can about the situation. "Listen, listen, listen." • Try to see the problem from the other person's perspective. • Be willing to be adaptable in seeking to be understood. • Present things logically, not emotionally. • Be credible, empathetic, and logical.
Habit 6: Synergize. **2 + 2 = 5**	• Make the whole greater than the sum of its parts. • Value the differences in the people you work with. • Foster open and honest communication. Help everyone bring out the best in everyone else.

Continues

The Seven Habits of Highly Effective People (*Continued*)

Habit 7: Practice Renewal: Sharpen the Saw. 	Renew the four dimensions of your nature: • Physical: exercise, nutrition, stress management • Mental: reading, thinking, visualizing, planning, writing • Spiritual: value clarification and commitment, study and meditation • Social/emotional: service, empathy, self-esteem, synergy

This table is meant to give only a thumbnail sketch of the seven habits. For a more complete discussion, see Covey's best-selling book, which includes numerous examples that illustrate these habits in action.

We continue our discussion of what constitutes a winning attitude by presenting advice for new graduates that was collected from company vice presidents, managers, and recent company hires. This advice was distilled and formulated into the list of seven actions necessary for a successful career.

The Seven Actions Necessary for a Successful Career

Action 1: Enjoy. 	• Find a job you enjoy. • Life is short: take time to reflect on what is important to you and make sure your career allows for those things. • Feel good about what you do or do something else. • Find time for health care. • Work hard, but have fun.
Action 2: Learn. 	• Continue to learn, renew, and expand your skill set through formal and informal training, reading, and conferences. • Never stop asking questions. Listen, question, and learn. Recognize what you know and what you don't know, and don't pretend to know what you don't. • Take advantage of other people's knowledge. Don't reinvent the wheel. • Learn how to take feedback, both positive and negative, and listen, listen, listen. • Learn how to interact with your colleagues, because the most effective results often come from the synergy created in effective team efforts, rather than as individual efforts.

Action 3: Communi-cate.	• Develop strong communication skills—oral, written, and listening. • The best work is of little value if you can't communicate it clearly and succinctly. • Learn how to market yourself and your results through sharing three key messages: what you do well, how you make a difference, and the kinds of challenges and projects you would like to take on. • Communication is a two-way street: develop active listening skills. • When you have something valuable to say, say it. • Develop a network of colleagues who can provide excellent advice and guidance. • Peers are great sounding boards.
Action 4: Work Hard.	• The harder you work, the better you'll do. • Focus on results. • Learn about the culture and politics of your organization. • Learn to manage up and down. Figure out what it takes to succeed. • Consider how your own strengths and ambitions align with the goals and aspirations of your job, organizations, and industry. • If the criteria for success in your organization are incompatible with your beliefs and your style, maybe that organization is not the right place for you.
Action 5: Evolve.	• Continue to learn, mature, and challenge yourself. Be prepared for changes in your career, and remember that every change is accompanied by new opportunities. • Find useful problems to work on. • Be willing to tackle different problems. • Life is constantly changing and your view of success will change with it so you will need to review it regularly and change direction as needed.

Continues

The Seven Actions Necessary for a Successful Career (*Continued*)

Action 6: Plan. 	• Be proactive in everything you do, especially in your career plans. • You own your career, so you need to develop your own picture of success. • Talk to people who are doing what you want to do 10 years from now to learn what experiences you will need to get there. • Pick a job (or a series of jobs) that meets your objectives. • Will they help you be where you want to be in 20 years from your career, financial, and personal points of view? • Find steps in the short term that take you in the direction you want to go in the long term.
Action 7: Share. 	• As your experience grows, share your knowledge with others. • Take time to find mentors and mentees. • Find a way to give something back to society.

Source: Supplemented with "Top Ten Tips for Taking Control of Your Career as a Leader," April 12, 2012, by SUPERADMIN. www.antoinetteoglethorpe.com/top-ten-tips-taking-control-of-career/.

Adopting and practicing the seven habits and seven actions will lead not only to better problem solving but also to your individual growth as a professional.

HAVING A VISION

"If you don't know where you are going, you'll probably end up somewhere else."
—Yogi Berra (New York Yankees, 1946–1965)

To make a difference, you must have a vision. A **vision** is the ability to see the way things ought to be or will be in the future. Each one of us must look to find the voids in our organization, our community, and our lives and then try to fill these voids. We can formulate a coherent, powerful vision by listening, reading, talking, and focusing our thoughts on bettering the current conditions. Our ethical and moral values help us establish our vision, which should be consistent with our personal belief system. Implementing a vision requires a master plan. A vision with a master plan makes day-to-day decisions easier by targeting our actions to achieve our desired outcomes.

To develop a vision, you must occasionally set aside a block of time (anywhere from a few minutes to several hours), become introspective, and step back and look at the "big picture." Sometimes we are so caught up with small details we lose sight of where we are headed and veer off course ("You can't see the forest for the trees"). Determine which direction your life (or your organization) should be taking, identify what needs to be accomplished, and devise a plan to meet your goals. Take a moment from time to time and reaffirm that the details you are working on are still aligned with the big picture.

Joel Barker, in his video *The Business of Paradigms*, speaks of the concepts of paradigm shifts, paradigm paralysis, and paradigm pioneers.[4] A **paradigm** is a model or pattern based on a set of rules that defines boundaries and specifies how to be successful at and within these boundaries. Success is measured in terms of the problems you solve using these rules.

Paradigm shifts can occur instantaneously, or they can evolve over a period of time. These transitions move us from seeing the world in one way to seeing it in an entirely different way. When a paradigm shifts, a new model based on a new set of rules replaces the old model. The new rules establish new boundaries and allow for solutions to problems that were previously deemed unsolvable. All practitioners of the old paradigm are returned to "the starting line" and are again on equal footing because the old rules no longer apply.

For example, the guidelines (rules) followed by the most successful manufacturer of slide rules became useless as a result of the paradigm shift in computation brought about by the invention of pocket calculators. Other more recent paradigm shifts include the ability to purchase products or services online. Traditional travel agencies have been forced to adapt and offer expanded services in order to compete with online services such as Travelocity and Orbitz. Other paradigm shifts have pitted DVDs against VHS tapes; online music sales against in-store sales; digital photography against film photography; dial-up modems against cable and DSL; MapQuest and Google maps and GPS units against AAA Trip-Tiks and gas-station maps; word processors against typewriters; online banking for standard recurring bills against paper checks each month; Google and online resources against encyclopedias; spell-checkers against dictionaries; and on-demand movies provided via the Internet against in-store rentals. With advances in technology now coming faster than ever, paradigm shifts can happen multiple times in the same industry. For example, floppy disks were the standard until their replacement by CD-ROMs in the 1990s, which were replaced by the USB drive in 2000, which is now being replaced by cloud computing.

Each of these technologies fundamentally changed the computing industry and returned all manufacturers to even ground.

In his second video, *Paradigm Pioneers*, Barker describes **paradigm paralysis** as a situation in which someone (or some organization) becomes frozen—that is, hopelessly locked into the idea that what was successful in the past will continue to

be successful in the future. **Paradigm pioneers** appear at the stage after a paradigm shift has occurred but it is not clear that the new paradigm will be successful. They are not pioneers in location but pioneers in time. They are people who have the courage to escape from paradigm paralysis by breaking the existing rules when success is not guaranteed. They realize that there are no easy roads when traveling in uncharted territory, so they cut new pathways, making the route safe and easy for others to follow. The characteristics of a paradigm pioneer are (1) the **intuition** to recognize a big idea, (2) the **courage** to move forward in the face of great risk, and (3) the **perseverance** to bring the idea to fruition.

Sukharevskyy Dmytro/
Shutterstock

Swiss Watch Paradigm Shift

In 1968, Switzerland, which had a long and highly respected history of making fine watches, accounted for approximately 80% of the world market in watch sales. Today, Swiss manufacturers hold less than 10% of this market. Their downfall: the emergence of the quartz digital watch.

You might be surprised to discover, however, that the Swiss invented the quartz digital watch. So why didn't these manufacturers capitalize on this invention? A paradigm shift in wristwatch technology had occurred. The Swiss failed to adopt the new technology because they were caught in a paradigm paralysis—the idea that what was successful in the past will continue to be successful in the future. After all, "the digital watch didn't have a mainspring, and it didn't tick; who would buy such a watch?" As a consequence of their paradigm paralysis, the inventors did not protect the digital watch with a patent, allowing Seiko of Japan and Texas Instruments (TI) to capitalize on the idea and market it. As a result of this miscalculation, the number of jobs in the Swiss watch industry dropped from about 65,000 to about 15,000 over a period of a little more than three years.

Of course, even if the Swiss had decided to manufacture the digital watch after realizing its success, they would have been simply on equal footing with Seiko and TI because of the paradigm shift. That is, all of the Swiss companies' vast experience in making watches with gears and mainsprings would have given them absolutely no advantage in manufacturing digital watches.

In the early 2000s the watch industry was turned on its head again as cell phones and other devices included clock functionality, eliminating the traditional utility of the watch. A paradigm shift in time-telling technology had occurred. After about five years of decreased watch sales, manufacturers created new value in watches by making them fashion accessories and including new functionality like temperature readings and news alerts, which again revitalized watch sales.

The Golden Arches Paradigm Shift

Ray Kroc was selling restaurant supplies in 1954, when he was surprised to receive a large order for multi-mixers from a small restaurant in San Bernadino, California. The restaurant, run by Dick and Mac McDonald, was unique in concept and very successful. It had a limited menu, consisting of hamburgers, French fries, and beverages. Rather than having many menu items, the McDonalds concentrated on quality for the few items that they served. Ray Kroc recognized the potential of this business strategy, which could be considered a paradigm shift from previous restaurant concepts. Kroc had the vision to recognize the value of this idea, and the courage to pursue the creation of a chain of McDonald's restaurants all over the country. In 1960, Ray Kroc bought the exclusive rights to McDonald's and the rest, as they say, is history. McDonald's has served billions of hamburgers over the years. Its limited menu and its attention to quality, service, cleanliness, and value in a chain of restaurants were true paradigm shifts for the industry.

Ray Kroc truly exhibited the three key characteristics of a paradigm pioneer: **intuition** (he saw an opportunity and realized the potential); **courage** (Kroc had the courage to push the idea of a fast-food restaurant focusing on quality, service, cleanliness, and value); and **perseverance** (he was able to persevere and bring the idea to fruition). As with all pioneers, many have since followed in his footsteps: Burger King, Wendy's, Taco Bell, and others.

www.mcdonalds.com

To be a paradigm pioneer you need not only to generate alternative solutions to a problem, but also to look for ways to improve things even when no apparent problems exist. Paradigm pioneers are continually searching for opportunities to initiate a paradigm shift to improve their process, product, organization, and so on.

Barker uses the example of the Swiss watch industry to make this point about paradigms.[4]

By now, the importance of paradigms, paradigm shifts, and having a vision should be obvious. New York Yankees catcher Yogi Berra best described the need to have a vision when he said, "If you don't know where you are going, you'll probably wind up someplace else." Once you have your vision of the future, it is important that you move forward with your vision. In Barker's third video, *The Power of Vision*, he says,

> Vision without action is merely a dream.
> Action without vision merely passes the time.
> Vision with action can change the world.

It is important not only to create a vision but also to periodically reevaluate your vision, modify it as necessary, and rework your implementation plan to accomplish it. Barker further discusses the importance of vision using a number of examples in his excellent video *The Power of Vision*.[4]

WORKING TOGETHER IN TEAMS

As problems become more complex and interdisciplinary in nature, their solutions tend to require the assembly of groups of people with different areas of expertise. These kinds of problem-solving activities will require your interaction with other people, either one-on-one or in group meetings.

The 1990s saw a dramatic increase in the use of teams in industry to formulate and solve problems. In the 2000s, many companies shifted to open offices, removing walls to create dynamic interactive environments. This shift away from individual problem solving and toward group-based processes came in response to the advent of global competition, which created a need to respond rapidly to changes in the market and changes in technology. As a result, universities were encouraged to give students more experience working in teams. In the material that follows, we have only enough time to give a thumbnail sketch of effective team problem solving. For a more in-depth study, you are referred to resources such as *The 17 Indisputable Laws of Teamwork*.[5]

As a member of a team, you will learn and practice collective decision making and collaboration. You will gain an appreciation of conflict and differences of opinion, learn to balance the time demands of the team with your other commitments, and, most of all, learn to appreciate the importance of mutual support.

Meetings are essential tools for team problem solving. They should be carefully planned and skillfully run to realize their maximum benefit. Because everyone's time is valuable, it is imperative that these meetings be both effective and efficient. Guidelines and sample meeting agendas as well as the Tuckman Model[6] for the Stages of Team Development can be found in the Professional Reference Shelf on the Web site.

The importance of meetings and positive group interactions cannot be over-emphasized. For the problem-solving process to function well, group members must learn to work smoothly together. All members of the team may not "like" all other members of the team, but they still must work together for the common good.

If you are working in a team environment, always be courteous, no matter how much you disagree with a team member. When disagreements occur, criticize the idea, not the person. Because mistakes will inevitably arise in the problem-solving process, focus on correcting these mistakes and preventing them from recurring, and then "move on." This focus on the issues forestalls defensive behavior and consequent loss of productivity. Provide positive reinforcement and encouragement to your team members. In many instances, the success of the project will depend on how well people communicate and interact with one another.

Another key ingredient of successful teams is the ability to work together creatively. John Sculley (the former chairman of Apple Computer, one of the more creative companies of the past decade) has discussed the philosophy of maintaining a creative environment for product development.[7] Such an environment will get people to reach beyond themselves and do extraordinary things. A couple of the ideas that he suggests for team leaders or managers to foster a creative environment are, first, tell your team where to go, not how to get there: roughly aim them in the direction at best. Creativity happens more easily when teams have the opportunity to think for themselves rather than live by another's directions. This freedom will allow teams to feel free to take their own way to a solution rather than worry about failing to follow specific directions, and, second, to safely raise the challenge. The workplace should be safe, so that workers are not afraid to take risks and make mistakes, but the standards should be set high.

As you become more involved in group problem solving, you will discover that personal traits inevitably surface among individuals within the group. Some traits will be positive; others will be a barrier to the group problem-solving process. When these challenges arise, the following table should help the group get through these issues and continue working toward a solution to the problem at hand.

How to Handle the Top Ten Problems in Groups[8]

Problem	Strategies to Minimize the Problem
Floundering	• Make sure the mission is clear and everyone understands what is needed to move forward.

Continues

How to Handle the Top Ten Problems in Groups (*Continued*)

Problem	Strategies to Minimize the Problem
Overbearing experts	• Have an agreement among team members that all members have the right to explore and question all areas. • Be courteous to everyone, no matter how they are behaving.
Dominating participants	• List "balance of participation" as a goal, and evaluate this issue regularly. • Practice "gatekeeping" to limit a dominant participant.
Reluctant participants	• Encourage everyone to participate. • Ask opinions of quiet members and encourage them by validation. • Require individual assignments and reports.
Rush to accomplishment	• Confront those doing the rushing, and remind them not to compromise quality for speed. • Strive to reach a balance of forward progress and decision quality by making sure all vital information is known and then making the best decision possible at that time. • Is there information not being considered that should be considered? If so, it is better to take the time to get the important details before moving forward without knowledge that could change decisions.

Unquestioned acceptance of opinion	• Play the devil's advocate; ask for supporting data and reasoning. • Accept and encourage conflicting ideas. • Be careful with criticism; criticize only ideas—not individuals.
Attribution of motives to others	• Ask for data to support statements. • Verify that the attribution is correct for a statement such as, "John is just saying this because he is angry and has an axe to grind with the Sales Department."
Discounting or ignoring a group member's statement	• Listening effectively is a must for all. Provide training in effective listening. • Support the discounted person. • Talk offline with anyone who continually discounts the opinions and contributions of other team members.
Wanderlust: digression and tangents	• Follow an agenda that includes time estimates. • Keep the topics in full view of the team, and direct the conversation back to the topic.

Continues

Feuding team members 	• Focus on ideas—not personalities. • Get adversaries to discuss the issues offline or get them to agree to a standard of behavior during meetings.

For a more complete discussion of ways to further minimize these top ten problems, see Peter Scholtes's work.[8]

RESPONDING TO CRITICISM

Responding to criticism in the workplace is a skill often lacked by employees, especially new hires. Excuses fly when people try to explain why things didn't get done on time, well, or at all. Bob Talbert, of the *Detroit Free Press*, researched and found the top ten most used excuses in the workplace.

Top Ten Most Used Excuses in the Workplace

1. I forgot.

2. No one told me to go ahead.

3. I didn't think it was that important.

4. I waited until the boss came back to ask him.

5. I didn't know you were in a hurry.

6. That's the way we've always done it.

7. That's not my department.

8. How was I to know that this was different?

9. I'm waiting for an OK.

10. That's his job—not mine.

Source: Bob Talbert, *Detroit Free Press*, Thursday, February 9, 1995.

Although avoiding these ten excuses altogether is nearly impossible, working to recognize when you are using one of them will help you be a better team member and a more efficient worker. High-achieving workers use criticism to learn and improve their performance while low-performing workers make excuses. The list is full of reasons why effort was not made to improve the status quo, act, or take responsibility. Instead, you should try taking the initiative to ask the boss rather than waiting for a reminder, challenging the way things are always done, or taking time to investigate small, seemly unimportant, details.

Safety Alert! No Brown M&Ms

Van Halen, a popular rock and roll band, arrived in Santa Fe, New Mexico, for a concert and immediately trashed their dressing room after finding brown M&Ms on their catering table. They also made sure the stage was checked for safety before performing that night. The band always requested M&Ms in their catering list, which they sent to venues in advance, and used an M&M clause as a contract reading test, which the Santa Fe venue failed.

The group was touring with 850 new lights at the time and they did not trust that setup crews at the various tour venues would be diligent in reading their contract, which detailed necessary safety procedures for setting up the stage with the new equipment.

In the middle of their extremely detailed performance contract the band randomly inserted the clause "There will be no brown M&M's in the backstage area or the promoter will forfeit the show at full price." So when they showed up and found the brown M&Ms, they knew the contract had not been read and they needed to do a serious safety check before performing (and could misbehave since they were contract-free that night). The band, upon performing a check of the stage, found that the stage had not been installed per the contract and was unsafe. In fact, the stage caused $470,000 worth of damage to the venue floor, for which the band was not liable.

It is likely that the venue staff didn't think reading the long detailed contract was that important (Excuse #3) or figured it was someone else's job (Excuse #10), and they installed the stage the way they always did with other stages (Excuse #6). It's also likely they failed to keep their jobs by using those excuses.

Source: www.npr.org/blogs/therecord/2012/02/14/146880432/the-truth-about-van-halen-and-those-brown-m-m-ms.

If you were an employee, would you have read the extremely detailed contract for the Van Halen concert? Maybe you did it the first time you were told to, but how about the 100th time when the first 99 had been the same? Imagine you had skipped reading the contract or breezed through it. Now your boss is extremely unhappy with you and threatening to fire you for allowing the venue to be damaged. How do you keep your job? How do you respond to criticism? Using an excuse is of no use.

Instead, how about adapt the following procedure that works in situations where a mistake has been made? Acknowledge that you made a mistake and state what you have learned from that mistake. Then, state the action you will take in the future to ensure that you will not make that mistake again. Be as specific as possible. "I made a mistake in not reading the contract thoroughly. I have learned that, although tedious, reading the contracts is very important. In the future, I will take the time to read all contracts so as to not miss any details." There is no guaranteed way to respond to criticism that will save you in every situation. Mistakes have consequences. Still, you will be better off if, instead of using excuses, you acknowledge your mistakes, learn from them, and take specific actions to avoid making the same mistakes in the future.

CONFLICT RESOLUTION

Controversy can be a positive contributor to the creative process; however, keeping controversy properly focused so that it does not degenerate into interpersonal conflict is a challenge for teams. Personality conflicts consume energy that would be much better spent on achieving team goals. Imagine now that you are the venue manager from the example with the brown M&Ms and your employees are pointing fingers at each other as to who is at fault for not reading the contract properly. Johnson and Johnson suggest the following negotiating steps to resolve a problem that arises from conflict between team members.[9]

Step 1: Describe Your Interests and What You Want
Tactfully describe your perception of the problem and what you want as a desired outcome.

- Define the conflict as small and specific, not general and global.
- Define your views in as short and as specific a manner as possible.
- Acknowledge the legitimate goals of the other person as part of the challenge to be addressed.
- Focus on a long-term cooperative relationship with statements such as "I think it's in the best interests of the team for us to talk about our argument."

- Be a good listener; face the other person and be quiet while that individual takes his or her turn.
- Show that you understand by paraphrasing what the other person said.

You could say, "It is clear that the contracts are not being read thoroughly. If we do not address this issue, we will likely make the same mistake again and be liable for more damages."

Step 2: Describe Your Feelings

Feelings must be openly expressed for the issue to be resolved. Acknowledging that every person's feelings are valid is essential for furthering the negotiation.

You might say, "I am confused why the contracts are not being read thoroughly."

Your employees might say they are frustrated about having to read the same contracts over and over. Or they may feel surprised and annoyed that one group had a different contract.

Step 3: Exchange Underlying Reasons for Your Opinions and Positions Relative to the Problem at Hand

It is now appropriate to better understand the underlying reasons both parties have for taking their positions.

- Present your reasons and listen to the reasons given by the other person. State the underlying reason(s) for what you want and work to understand the other party's reasons. Only through this empathetic understanding can you search for creative, win/win solutions.

You could then say, "As manager it is my priority that we run a safe and profitable concert venue, and maintaining proper knowledge of our contracts is important to that priority."

Step 4: Understand the Other Person's Perspective

Clarifying the intentions of your teammate may help you realize that his or her intentions are not the same as your fears. Be sure that you understand both perspectives, and openly discuss opposing perceptions.

Some Tips for Effective Listening
- Make eye contact.
- Avoid negative body language (e.g., constantly looking at your watch).
- Practice "active listening" by using encouraging verbal cues to elicit further information (e.g., "Can you expand on that?" or "Tell me more").

- Frequently confirm your understanding by restating what you think you have heard others say.
- Use Covey's fifth habit of highly effective people: "Seek first to understand, then to be understood."

Lastly, you could say, "I see that it would be frustrating to have to read the same contract over and over. However, that is not enough of a reason to not have them properly read."

Step 5: Generate Options for Mutual Gain

Use each other's perspective to promote the generation of new, creative solutions. Empower the other person by being flexible and providing a choice of options. Brainstorm to generate at least three workable alternative agreements before selecting the one solution that you will jointly employ. Use the techniques discussed in Chapters 6 and 7 to help generate options.

Allow everyone to participate in this process: you could standardize contracts, read them yourself, or take turns reading them instead of having one person do it all the time.

Step 6: Reach a Wise Agreement

- Does everyone have an equal chance of benefiting?
- Does the agreement meet the legitimate needs of everyone in the team (or at least those people who are directly affected by the conflict)?
- Are the gains and losses of all parties roughly in balance?

As a manager, you have the opportunity to fix the underlying problem while making sure things run smoothly while a new solution is found and implemented. "I will look into standardization of all contracts so we have to read less for each event. In the meantime, I am going to assign a specific person who will be held accountable for reading all contracts and knowing every detail so the next time a situation like this comes up we catch the details."

SUMMARY

This chapter began by emphasizing the importance of approaching the problem with a positive, "can-do" attitude, striving to develop the traits of expert problem solvers, and practicing Covey's seven habits of highly effective people. We continued with a discussion of the need to have a vision and to look for paradigm shifts and exploit them. Finally, we presented a micro-summary of ideas for working effectively in teams to solve problems.

- **Practice**
  - Recognize and cultivate the characteristics of expert problem solvers.
  - Practice the seven habits of highly effective people.
  - Practice the seven actions for a successful career.

- **Paradigms**
  - Welcome change and paradigm shifts as opportunities to make inroads and advancements.
  - A **paradigm** is a model or pattern based on a set of rules that defines boundaries and specifies how to be successful within these boundaries.
  - **Paradigm shifts** occur when a new model based on a new set of rules replaces the old rules. We establish new boundaries, and we allow more problems to be solved.
  - **Paradigm paralysis** occurs when someone or some organization is stuck on the idea that what was successful in the past will be successful in the future.
  - **Paradigm pioneers** are individuals who have the courage to move forward and to break existing rules when success is not guaranteed.

- **Vision**
  - Look for and fill the voids in your organization. Try to provide a vision to fill those holes and move the organization forward.
  - Develop an atmosphere that encourages and fosters creativity in your co-workers.
  - Listen to your customers and work with them as a unit to develop creative solutions.

- **Teamwork**
  - Seek ways to establish a creative team environment.
  - Follow the suggestions on ways to handle the top ten group problems.
  - Pursue conflict resolution to ensure that your team will work effectively and efficiently.

WEB-SITE MATERIAL (WWW.UMICH.EDU/~SCPS)

- **Learning Resources**
  Summary Notes
  Closed-Ended Algorithm (accessible from the home page)
  PowerPoint Slides

LIVERPOOL JOHN MOORES UNIVERSITY
LEARNING SERVICES

- **Interactive Computer Module Introduction**

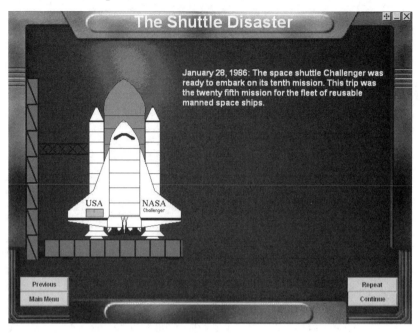

Introduction

- **Professional Reference Shelf**
  1. Characteristics of Effective and Ineffective Problem Solvers
     - Attitude
     - Actions
     - Solution Procedures
     - Accuracy
  2. Closed-Ended Problem Algorithm on How to Guess the Number of
     - Jellybeans in a Jar

  3. The Tuckman Model for the Stages of Team Development
     - Forming stage
     - Storming stage
     - Norming stage
     - Performing stage
  4. Establishing a Creative Team Environment
     - Goals

     - Contrarian thinking
     - Inspire
     - Emotion
     - Encourage

5. Top Ten Lists
   – Top Ten Faults of Executives
   – The Ten Most Used Excuses in the Workplace
   – Ten Commandments for Losing
6. The Key Is Listening to Your Customers
7. Guidelines for Running Effective Meetings, Including a Typical Agenda for Meetings

REFERENCES

1. Whimbey, A., and J. Lockhead, *Problem Solving and Comprehension: A Short Course in Analytical Reasoning*, Franklin Institute Press, Philadelphia, 1980.
2. Wankat, P. C., and F. S. Oreovicz, *Teaching Engineering*, McGraw-Hill, New York, 1992.
3. Covey, Stephen R., *The Seven Habits of Highly Effective People*. © 1989. Reprinted with permission of Franklin & Covey, New York, 1989.
4. Barker, J. A., *Discovering the Future: The Business of Paradigms*, ILI Press, St. Paul, MN, 1985; *The New Business of Paradigms, 21st Century Editions*, distributed by Star Thrower (800-242-3220); *Paradigm Pioneers and the Power of Vision, 21st Century Editions*, distributed by Star Thrower (800-242-3220). Also see http://abcnews.go.com/Technology/story? id=1518077#. TzmFJhy feME and www.npr.org/2010/11/08/131163403/its-time-the-wristwatch-makes-a-comeback.
5. Maxwell, John C., *The 17 Indisputable Laws of Teamwork*, Thomas Nelson, Nashville, TN, 2001.
6. Tuckman, Bruce W., "Developmental Sequence in Small Groups," *Psychological Bulletin*, 63, pp. 384–399, 1965. The article was reprinted in *Group Facilitation: A Research and Applications Journal*, 3, Spring 2001, http://dennislearningcenter.osu.edu/references/GROUP%20DEV%20ARTICLE.doc, accessed January 14, 2005. Also see Hanwit, Jessie, *Four Stages of Teambuilding*, http://education.wm.edu/centers/ttac/resources/articles/transition/fourstageteam/index.php.
7. Sculley, John, and John A. Byrne, *Odyssey: Pepsi to Apple... A Journey of Advertising Ideas and the Future.* © 1987 by John Sculley. Reprinted by permission of HarperCollins Publishers, Inc.
8. Scholtes, Peter R., *The Team Handbook*, Oriel, Inc. (formerly Joiner Associates, Inc.), Madison, WI, 1988.
9. Johnson, David W., and Frank P. Johnson, *Joining Together: Group Theory and Group Skills*, 9th ed., Allyn and Bacon, Boston, MA, 2005.

EXERCISES

2.1. Load and run the "Interactive Computer Module Introduction" from the Web site. At the end of the scenario, a performance number will appear on the screen. Record the performance number (optional: turn it in to your instructor). This number will let you know how well you did on this exercise. Performance number = _____.

2.2. Review "Characteristics of Effective and Ineffective Problem Solvers" on the Web site. Which characteristics listed in the table do you feel you now possess? Which characteristics do you feel you need to improve?

2.3. Review the seven habits of highly effective people.

    A. Explain how you will practice at least four of the seven habits during the coming three weeks.

    B. Make a list of at least three habits that you observed others using or practicing.

2.4. Discuss which of the seven actions for a successful career you believe are the most important, and explain why they are important to you. Describe how you will practice the actions most important to you. Make a list of at least three actions that you observed others using or practicing.

2.5. Make a list of at least three other examples of recent paradigm shifts you can identify.

2.6. Look around an organization of which you are a member for things that could be improved upon. Make a list of the voids that need to be filled to make the organization even better and more effective. Which of these changes would really alter the way the organization functions? What would need to be accomplished to produce a paradigm shift? List two ways you can practice being a paradigm pioneer.

2.7. Identify a group of people with whom you frequently interact.

    A. List four things you can do to become a better team member and four ways for you to establish a creative environment in your group.

    B. Think of four adjectives that describe yourself. How do each of these adjectives make you a better/worse team member?

2.8. You are in a group of four working as a team to define and solve a problem.

    A. Discuss the Tuckman model described on the Web site and how you will use it to guide your group's activities.

    B. Describe how you would handle each of the following situations:

        1. Someone starts to dominate the group discussion and direction.

        2. Two of the group members are good friends and seem to form a clique.

        3. One member of the group is not carrying his or her load.

        4. One member of the group continually makes mistakes in his or her part of the project.

2.9. Review "Establishing a Creative Environment" on the Web site. Prepare a list of at least four specific ideas that would establish a creative environment in your group.

2.10. After reviewing the "Guidelines for Running Effective Meetings" in the Professional Reference Shelf on the Web site, develop an agenda for the first meeting involving the following groups:

    A. Your colleagues, to write a report for an undergraduate laboratory course

    B. The floor of your residence hall, to inform the new students of the rules, traditions, and other operations of the residence hall

    C. A local interest group you are to lead

2.11. An instructor requested that students in the class form six-person teams, attempting to maximize diversity and selecting people to work with who are new to them. A team was formed that was composed of two white men, two white women, one African American man, and one Asian American man. The group selected one of the white men as their team leader. During the first and second meetings of this team, which took place during class time and in the classroom, the instructor noted that the African American man and the white man who was not the team leader sat almost outside the circle formed by the other team participants. Moreover, the white man who was the team leader and one of the white women appeared to do all the talking and to make all the suggestions about how to proceed. The other four people on the team looked uninvolved, at least as far as the instructor could observe.

  A. What might be going on in this team?

  B. Should the instructor intervene? Why? How? What would you do?

  C. Could any of the steps of conflict resolution be applied here?

  D. What preparation, training, or instruction for teamwork might have helped this team? What training or instruction might be helpful to it now?

  E. What preparation or instruction in teamwork dynamics, supervision, or intervention might be helpful to you in this and similar situations?

  This exercise is adapted from the FAIRTEACH Workshop with the University of Michigan's School of Engineering Faculty, Martin Luther King, Jr., Day, 1994.

2.12. Review the exercise (in the Professional Reference Shelf on the Web site) on how to guess the number of jellybeans in a jar. Redo the jellybean contest guessing problem for cylindrical gourmet jellybeans in a one-gallon cylindrical vessel.

2.13. Choose your own criteria and then pick the two best items from each of the three top ten lists in the Professional Reference Shelf on the Web site.

2.14. Go to www.umich.edu/~elements/asyLearn/itresources.htm.

  A. Take the "Inventory of Learning Style" test, and record your learning style according to the Solomon/Felder inventory:

  – Global/Sequential

  – Active/Reflective

  – Visual/Verbal

  – Sensing/Intuitive

  B. After checking on www.umich.edu/~elements/asyLearn/itresources.htm, suggest two ways to facilitate your learning style in each of the four Felder–Soloman learning style categories.

  C. Visit the problem-solving Web site at www.umich.edu/~elements/probsolv/closed/cep.htm to find ways to "get unstuck" when you get stuck on a problem and review the "Problem-Solving Algorithm." List four techniques that might help you in finding solutions to the homework problems.

FURTHER READING

Gunneson, Alvin. "Communicating Up and Down the Parks," *Chemical Engineering*, 98, p. 135, June 1991. Useful tips on how to improve your interactions with those employees above, at the same level, and below you in your organization.

Lumsdaine, E., and M. Lumsdaine. *Creative Problem Solving: An Introductory Course for Engineering Students.* McGraw-Hill, New York, 1990.

Mathes, J., and D. Stevenson. *Designing Technical Reports*, 2nd ed. New York, Simon & Schuster, 1991.

Phillips, Denise A., and A. E. Ladin Moore. "12 Commandments." *Chemtech*, 21, p. 138, March 1991. Rules to help improve your communication skills.

Raudsepp, Eugene. "Profits of the Effective Manager." *Chemical Engineering*, 85, p. 141, March 27, 1978. Although this article was written 15 years ago, these traits still apply to effective leadership.

Strunk, W., and E. B. White. *The Elements of Style*, 4th ed. Macmillan, New York, 2000. A concise treatise on grammar rules and writing style with many examples.

VanGundy, Arthur B., Jr. *Techniques of Structured Problem Solving*, 2nd ed. Van Nostrand Reinhold, New York, 1988.

Whimbey, A., and J. Lockhead. *Problem Solving and Comprehension: A Short Course in Analytical Reasoning*, 2nd ed. Franklin Institute Press, Philadelphia, 1980.

3   SKILLS NECESSARY FOR EFFECTIVE PROBLEM SOLVING

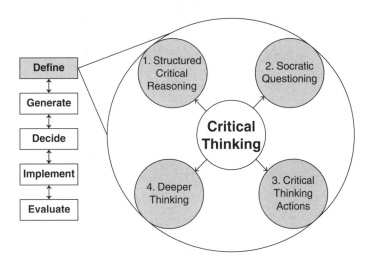

CRITICAL THINKING

Critical thinking is the process we use to recognize underlying assumptions, scrutinize arguments, question problem statements and solutions, and interpret and assess the accuracy of information. Critical thinking involves objectivity, analysis, evaluation, and drawing conclusions in a structured and well-reasoned way. Critical thinkers are persistent in their search for evidence and implications of a viewpoint, and they evaluate the strengths and weaknesses of the evidence. They continually ask probing questions of themselves and others.

Critical thinking is one of the most important skills you can possess and is vital to good problem solving. This skill is applicable in everything you do, whether it's related to work, friends, family, or any other area of your life. Critical thinking can help you to define and solve real problems, to ask the right questions, to decide if a proposition or solution is valid, or to suggest a path forward for an important issue.

There are a number of great books devoted entirely to critical thinking, which we cannot possibly distill here. What we do hope to do is to provide a few central fundamental ideas and useful techniques and exercises that you can use to develop and practice critical thinking skills. The two areas we will focus on are (1) **structured critical reasoning (SCR)**, a critical thinking algorithm used to analyze a document, proposition, or problem solution, and (2) **Socratic questioning**, a way to ask the right questions in order to distinguish the real problem from the stated or perceived problem.

Structured Critical Reasoning

The algorithm we will use to analyze a proposition, thesis, and so on is sometimes called structured critical reasoning (SCR) and has been used to unravel even the most complex arguments.

The sequence of the SCR analysis is to identify the following:

- Conclusions
- Evidence
- Assumptions
- Strengths and weaknesses of each assumption
- Fallacies in logic

The SCR algorithm for this sequence is based on the work of Browne and Keeley[1] and is expanded in the following table.

Structured Critical Reasoning

Step 1. Identify all of the author's conclusions.

A conclusion is a statement or idea in a document or speech that the writer or speaker wants you to accept. Make a list of all the conclusions in the document/proposition/presentation. When looking for the conclusion, ask yourself first "What are the issues?" To rapidly identify the conclusion, Browne and Keeley[1] suggest looking for indicator words such as *therefore, consequently, which leads us to, proves that, the point is*, and so on in the written statement or presentation you are given.

Step 2. Look for the reasons and evidence the author uses to support each conclusion.

There is an important distinction between reason and evidence.

Reasons are *internal evaluations* that can be based on facts and data but are not necessarily well substantiated. Many times, reasons are based on feelings, personal experiences and observations, intuitions, or beliefs such as "I think this statement is true because …" Reasons are often put forth as evidence and it is up to the analyzer to decide if they are valid.

Evidence is based on *external evaluations*, such as facts, data, laws, observations, case examples, or research findings. All evidence are reasons, but not all reasons are evidence.

For each conclusion make a list of all evidence that has been given that you think supports the conclusion. How strong is each piece of evidence? Does the evidence support the conclusion? What evidence would cause you to reject the conclusion? Is there a general lack of evidence or has significant information been omitted?

Step 3. List all major assumptions.

An assumption is a belief we use to support the evidence. Make a list of the assumptions in each piece of evidence. Look for hidden or unspoken assumptions (e.g., "A company designs a new pencil that will stay sharper much longer than all competing pencils so they project big sales in the first year"). The assumptions might be that customers will want to buy the pencil just because it stays sharper longer than other pencils; that the competition is not also launching a new pencil that will stay sharper longer; and that demand for pencils will not drastically fall in the next year. A hidden assumption is that the new erasable ink pens will not affect the market for the new pencils.

An employee reported to his supervisor that his work team was not functioning well. He spoke generally about friction between members of the team. The supervisor stated that she would look into it. She noted that just prior to the complaint a new member had been added to the team. Her hidden assumption was that, since the complaint and the new member's arrival coincided, there must be a connection. She transferred the new member to a different team and was surprised when the workgroup continued to have friction and communication problems.

Step 4. Evaluate all the assumptions and evidence.

Each assumption must be evaluated to determine whether it is strong or weak and whether it is relevant to the conclusion. If assumptions are irrelevant, or contain contradictions, and/or contain fallacies they likely do not provide support for the conclusion. All assumptions are hypotheses, and it is up to the evaluator to put forward his or her best judgment as to whether or not the assumptions are good or questionable. If you become stuck evaluating an assumption, list all the pros and cons for accepting the hypotheses and then make a decision. A balance must be struck between scrutinizing assumptions and making progress in the analysis.

Step 5. Identify fallacies in logic.

The following table gives 11 common fallacies to look for when evaluating the assumptions used in supporting the evidence and the conclusions. In some instances more than one fallacy can apply to the situation.

Eleven Fallacies in Logic to Look For

1. **Ambiguous or vague words or phrases:** Uses words, phrases, or sentences that have multiple interpretations or really don't say anything.

   "The model is in close agreement with the data." What does the word "close" mean? What is the measure of a *"close agreement"*? Within 10%? 50%?

2. **Citing a questionable authority:** Gives credibility to someone who has no expertise in the area.

   John agrees with me that drinking energy drinks are bad for you. What makes John an expert on the perils of drinking energy drinks? John could be an expert dietician studying the subject or he may have no basis for knowing anything about the effects of energy drinks on the body other than an uneducated opinion.

3. **Straw person:** Discredits an exaggerated version of an argument.

   Recent auto accidents in your neighborhood have led you to propose to the city council that the speed limit along Main Street be reduced to calm the traffic flow. Opponents complain that reducing speed limits all over town is counterproductive and an unnecessary burden on drivers.

   The straw person argument here is the expansion of your proposal from "a lower speed limit on one street" to "speed limits all over town." The acknowledgment of this new alternative argument deflects the focus from your true proposal.

4. **False dilemma (the either-or):** Assumes the choices stated by the author are the only ones that exist, or generally constrains the scope of a discussion to force a point.

   At a recent cocktail party, the conversation has turned to family pets, and your friend asks you "Are you a cat or a dog person?" Your choices here have clearly been limited to two, when in reality there are many others: you may have no interest in pets at all, you may be a bird person, or you may enjoy cats and dogs equally.

5. **Red herring:** Introduces an irrelevant topic to distract the conversation from the main point.

   You call your cell phone provider to complain about how poor your cell phone battery life is after the recent software update. The representative, instead of responding to your concern, praises the provider's new unlimited text-messaging plans that are due to be released in the next month.

Your phone's battery life will not be improved by being able to send more text messages.

6. **Slippery slope:** Assumes that if this fact is true then everything else follows.

   A father talking to his daughter on dating a boyfriend he doesn't like says, "If you continue dating this guy who doesn't take his education seriously, you'll end up dropping out of school, you then won't be able to get a job, and you will get married too young."

   Dating someone who doesn't take education seriously does not mean the daughter will drop out of school herself, marry early, and be unemployable.

7. **Appeals to popularity:** Justifies an assumption by stating that large groups have the same concern or that anything favored by a large number of people is desirable.

   An opinion article in a campus newspaper states that in an all-campus survey 95% of students think that tuition should be lowered and therefore tuition should be lowered immediately.

   The students are biased because they have to pay tuition and are not inclined to think of the budget problems that would be caused if the school lowered tuition for all students.

8. **A "perfect" solution:** Assumes that if a part of the problem is not satisfied or solved (even a small part), then the entire solution should be abandoned.

   "Don't waste your money on a home security system; master thieves will still be able to get into your house."

   However, many thieves may be deterred by a security system.

9. **False, incomplete, or misleading facts or statements:** Presents data in such a way that it falsely leads someone to the wrong conclusion.

   "Because 90% of college students polled had no debt, education costs are not a problem."

   It's possible that only 10 college students were polled or the poll was taken at a banquet for scholarship students.

10. **Causal oversimplifications:** Explains an event by attributing it to a single factor, when many factors are involved or by overemphasizing the importance of a single factor.

Continues

At a party you overhear a friend tell her spouse, "I had high blood pressure at the doctor's office today; I really need to reduce the stress in my job."

This friend is obviously attributing the high blood pressure reading to job-related stress, while there may be many additional contributing or more important factors (lack of exercise, poor diet, genetic predisposition, white-coat syndrome, etc.).

11. **Hasty generalizations:** Draws a conclusion about a large group based on the experiences of a few members of the group.

    All engineers are introverts who would rather relate to computers than people. All football players are dumb jocks.

    Clearly there are many engineers who are outgoing and football players who are very intelligent. It is very dangerous to make sweeping generalizations regarding a group based on limited experience.

Bias and Lack of Information

The structured critical reasoning (SCR) heuristic offers a solid foundation upon which to deconstruct *presented* arguments for validity. However, it is also important to recognize what is not presented. Often individuals will omit significant information about an argument in order to make the answer overwhelmingly clear. This can be done because the individual is biased to one side of an argument and knowingly only presents supporting evidence or because that person has a general lack of knowledge of the argument. Regardless of cause, lack of information is important to be aware of as you apply the SCR algorithm. A few ways to check for bias and lack of information are to ask what evidence you think should be required to support a given conclusion, to look into the author's background, or to find information about the topic from a variety of sources to see what evidence is presented.

Now, let's apply SCR. We begin with a confused mayor in California.

Fighting Fires?

In Orange County, California, only 2% of firefighter emergency responses involve fires; the rest are car accidents, fender benders, bicycle accidents, and other small medical emergencies. This unnecessary deployment of firefighters wastes money by sending gas-guzzling fire trucks and full fire crews to situations where they are not needed. The mayor responds to the data by stating that because firefighters are out at the streets on nonfire emergencies, they may not be available to respond to fires in the county. He recommends that new fire stations and detection systems be implemented throughout the region to be available to respond to county fires.

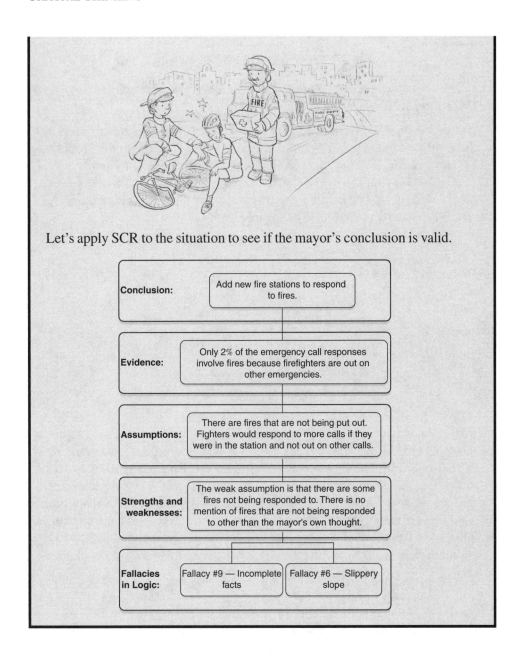

Let's apply SCR to the situation to see if the mayor's conclusion is valid.

Conclusion:	Add new fire stations to respond to fires.
Evidence:	Only 2% of the emergency call responses involve fires because firefighters are out on other emergencies.
Assumptions:	There are fires that are not being put out. Fighters would respond to more calls if they were in the station and not out on other calls.
Strengths and weaknesses:	The weak assumption is that there are some fires not being responded to. There is no mention of fires that are not being responded to other than the mayor's own thought.
Fallacies in Logic:	Fallacy #9 — Incomplete facts Fallacy #6 — Slippery slope

It is clear that the mayor's conclusion to add new fire stations is not well supported because the evidence relies on a weak assumption with fallacies in logic. The mayor's solution is Fallacy #6 (slippery slope): if we were to believe that there are fires that are not being responded to, then his solution of more fire stations makes sense. There is incomplete information (Fallacy #9) here as the mayor does not know that there are fires not being responded to. As is sometimes the case, we find two fallacies in logic in the mayor's conclusion. We continue with a more complex example examining an opinion article from a London newspaper.

SCR: Truancy in U.K. Schools

Carry out an SCR analysis of the following synopsis of an article written by a British teacher with 20 years of experience.

Help recession-hit families to decrease truancy: Soaring truancy rates in the United Kingdom are not surprising to teachers like me. Figures from the Department for Children, Schools, and Families show that children skipped more than eight million days of school last year. There are many reasons for the rising number of truants, but I think there is one big underlying reason: the recession is really beginning to bite in many households.

In the United Kingdom, four million children live below the poverty line, and that number is rising. Charities, such as Save the Children, are seeing families of four trying to feed themselves on 20 to 25 British pounds a week. That means that many children are living in households under severe stress, frequently working illegally or carrying out household chores for parents who need them at home. One student I taught some time ago wound up spending quite a few days at home taking care of her younger brother and sister while her mom went out to work. She skipped school at the insistence of her mother.

The statistics show that these cases are more and more common and now, unlike in previous years, increasing numbers of parents are being jailed for having truant children. Ministry of Justice figures released this year reveal that the number of court-issued penalty notices went up by 12% last year to 7,793. It is clear that families are being torn apart by truancy.

Rather than addressing the root causes of truancy, the government is too keen to criminalize desperate parents. Work by charities such as Save the Children shows that when needy families with truant students are helped properly, the truancy issue can be solved much more cheaply and wisely than by incarcerating a child's main caregiver. Proper investment in public services for recession-hit families is vital for decreasing truancy rates.

Source: www.guardian.co.uk/commentisfree/2009/oct/21/education-bullying-truancy-recession-care.

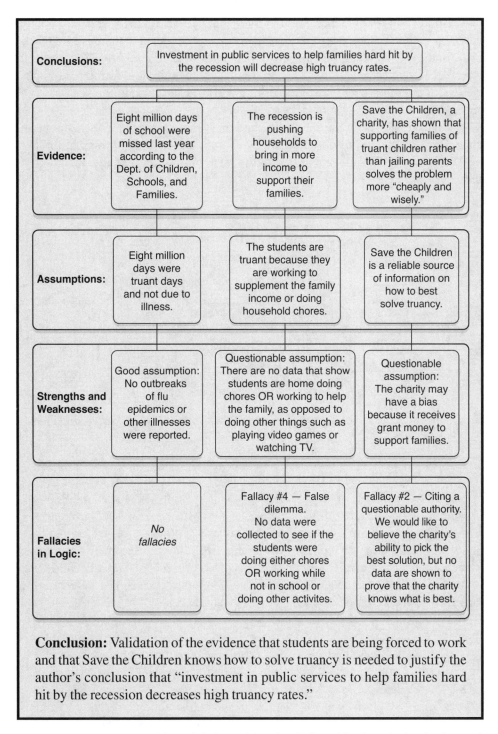

Conclusions:	Investment in public services to help families hard hit by the recession will decrease high truancy rates.		
Evidence:	Eight million days of school were missed last year according to the Dept. of Children, Schools, and Families.	The recession is pushing households to bring in more income to support their families.	Save the Children, a charity, has shown that supporting families of truant children rather than jailing parents solves the problem more "cheaply and wisely."
Assumptions:	Eight million days were truant days and not due to illness.	The students are truant because they are working to supplement the family income or doing household chores.	Save the Children is a reliable source of information on how to best solve truancy.
Strengths and Weaknesses:	Good assumption: No outbreaks of flu epidemics or other illnesses were reported.	Questionable assumption: There are no data that show students are home doing chores OR working to help the family, as opposed to doing other things such as playing video games or watching TV.	Questionable assumption: The charity may have a bias because it receives grant money to support families.
Fallacies in Logic:	No fallacies	Fallacy #4 — False dilemma. No data were collected to see if the students were doing either chores OR working while not in school or doing other activites.	Fallacy #2 — Citing a questionable authority. We would like to believe the charity's ability to pick the best solution, but no data are shown to prove that the charity knows what is best.

Conclusion: Validation of the evidence that students are being forced to work and that Save the Children knows how to solve truancy is needed to justify the author's conclusion that "investment in public services to help families hard hit by the recession decreases high truancy rates."

The conclusion of this article is the idea that is found both at the beginning and end of the author's statement: Investment in public services to help families who are hard hit by the recession will decrease high truancy rates.

By looking for evidence, we find three main points. In the first paragraph the author says that truancy rates are soaring to convey the idea that there is a lot of truancy. In the second paragraph the idea is presented that students are being forced to work because of the recession, causing truancy. In the third and fourth paragraphs, we read that jailing parents is not working and instead we should invest in social services.

On page 49 we put the conclusion and these three pieces of evidence in flowchart form and continue our analysis by examining the assumptions in the evidence, the strengths and weaknesses of the assumptions, and, finally, the fallacies in logic.

SCR: A Public Health Hazard—Eggs[2]

Carry out an SCR analysis of the following synopsis of some articles written about a recent study examining the health effects of eggs.

A recent study performed by Canadian medical researchers on the health effects of eggs has caused quite a stir. They compared the cardiovascular risks associated with eggs to that of smoking. This led to a series of news reports with sensationalized titles like "Eggs Are Nearly as Bad for Your Arteries as Cigarettes" and "Are Eggs the New Cigarettes?" The study involved approximately 1,200 subjects about equally split between men and women who were being treated for cardiovascular diseases. The average age was 61. On their first visit to the Canadian vascular prevention clinics, the subjects were surveyed for some baseline characteristics, including blood cholesterol, blood pressure, and body mass index, and their total carotid plaque area (mm^2), TPA, was measured ultrasonically. Personal habits were also tabulated with a lifestyle survey at the initial visit. Egg consumption and smoking behavior were estimated by the subjects. For egg consumption, if a subject said he or she consumed two eggs per week for the past 50 years, a "score" of 100 egg-yolk years was given. Similarly, smoking was estimated by the number of packs per day times the number of years the individual was a smoker (30 years as a smoker of 0.5 packs per day = 15 pack-years). Alcohol consumption and exercise were not taken into account because the textual responses were too hard to quantify ("quit drinking six years ago" and "plays golf twice a week"). The study concluded that the effect of egg consumption was approximately two-thirds of the deleterious effect produced by smoking on cardiovascular health because the TPA increased for egg-yolk years at two-thirds of the rate it did for pack-years. Interestingly, the group with the highest egg-consumption (average age, 69.77; egg-yolk years greater than 200 years) had the lowest total cholesterol and the lowest body mass index but the highest TPA of all the groups surveyed.

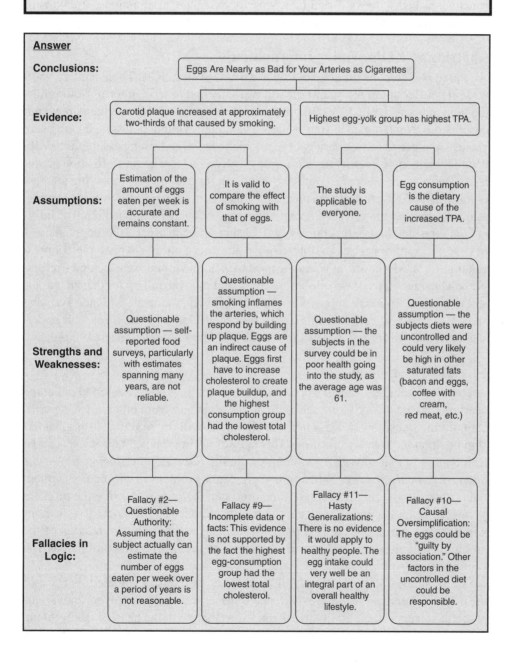

Answer

Conclusions:

Eggs Are Nearly as Bad for Your Arteries as Cigarettes

Evidence:

Carotid plaque increased at approximately two-thirds of that caused by smoking.

Highest egg-yolk group has highest TPA.

Assumptions:

Estimation of the amount of eggs eaten per week is accurate and remains constant.

It is valid to compare the effect of smoking with that of eggs.

The study is applicable to everyone.

Egg consumption is the dietary cause of the increased TPA.

Strengths and Weaknesses:

Questionable assumption — self-reported food surveys, particularly with estimates spanning many years, are not reliable.

Questionable assumption — smoking inflames the arteries, which respond by building up plaque. Eggs are an indirect cause of plaque. Eggs first have to increase cholesterol to create plaque buildup, and the highest consumption group had the lowest total cholesterol.

Questionable assumption — the subjects in the survey could be in poor health going into the study, as the average age was 61.

Questionable assumption — the subjects diets were uncontrolled and could very likely be high in other saturated fats (bacon and eggs, coffee with cream, red meat, etc.)

Fallacies in Logic:

Fallacy #2— Questionable Authority: Assuming that the subject actually can estimate the number of eggs eaten per week over a period of years is not reasonable.

Fallacy #9— Incomplete data or facts: This evidence is not supported by the fact the highest egg-consumption group had the lowest total cholesterol.

Fallacy #11— Hasty Generalizations: There is no evidence it would apply to healthy people. The egg intake could very well be an integral part of an overall healthy lifestyle.

Fallacy #10— Causal Oversimplification: The eggs could be "guilty by association." Other factors in the uncontrolled diet could be responsible.

Applying SCR: Examining Information

The previous examples have demonstrated how to deconstruct a conclusion presented to you in written form, and the same process applies when you listen to an argument. You simply process the information as it is being said rather than reading. It is often advantageous to take notes and construct an SCR analysis during a presentation in order to find weak assumptions and fallacies in logic. You can practice this by watching TV pundits speak on their views. Ask yourself, "What are the conclusions being drawn?" and then look for evidence, assumptions, strengths of assumptions, and fallacies in logic.

Applying SCR: Presenting Information

We move now to constructing your own positions using SCR. There are two ways we will consider presenting information: verbally and in written form. Both follow the same process. Begin by examining all information available and then draw your conclusions. Next, organize your evidence to support your conclusion, making sure you use referenced facts, a variety of reputable sources, and strong assumptions with no fallacies in logic in order to gain validity. Once you have your information organized, you can communicate it orally or in written form simply by following the flow of information you created with the SCR. This process can be used in essay writing, sales, debate, and more. When open to bias as previously mentioned (lack of information, one-sidedness), this process will ensure you construct a thorough analysis of your positions. It is also possible to move this information into paragraph or presentation form. Moreover, SCR can be used to examine topics on which no irrefutable conclusion can be drawn. Policy, politics, and more are all up for debate but it is still possible to choose a strong position with SCR. Examine all evidence available and draw your best conclusion. Even if your SCR is not flawless, you will understand the areas an argument is lacking: weak assumptions and xfallacies in logic.

Socratic Questioning

Asking the right questions in a presentation, meeting, or conversation to get at the heart of an issue is a skill that sets critical thinkers apart from others. Asking critical thinking questions (CTQs) in these situations will put you in a strong leadership position in your organization. This skill of asking the right questions can be learned and practiced with Socratic questioning. *Socratic questioning lies at the heart of critical thinking.* When you are given a problem or problem statement rather than discovering it yourself, it is important that you make sure the problem you were given accurately reflects the true situation. Asking Socratic questions will help you ferret out the real problem. It helps identify the boundaries of the problem and helps you learn if you are getting to the heart of the problem as you continue to question.

Our studies on problem-solving techniques in industry revealed that one of the major differences between experienced, successful problem solvers and novice problem solvers is their ability to ask questions that go to the heart of the problem. Experienced solvers tend to interview as many people as necessary that might possess useful information about the problem and to use critical thinking to reflect on, assess, and judge the assumptions underlying the information they collect. We will use R. W. Paul's six types of Socratic questions[3] to explore the proposed problem statement and/or a question that has been asked. While many types of Socratic questions exist, we have selected six types to apply in the following critical thinking questions (CTQs) examples shown in the right-hand side of the table. For a more complete listing, refer to the Web site's Summary Notes for this chapter.

Six Types of Socratic Questions and Examples of CTQs

1. Questions about the question or problem statement The purpose of this question is to find out why the question was asked, who asked it, and why the question or problem needs to be solved.	• What was the point of this question? • Why do you think I asked this question? • Why is it important you learn the answer to that question? • How does that question relate to our discussion? • Where did the problem originate?
2. Questions for clarification The purpose of this question is to find missing or unclear information in the problem statement question; identify multiple interpretations and ambiguous words and phrases.	• What do you mean by _____? • Why do you say that? • What do we already know about that? • Could you explain further? • Could you put that another way?
3. Questions that probe assumptions The purpose of this question is to find out if there are any hidden, misleading, or false assumptions.	• What could we assume instead? • How can you verify or disapprove that assumption? • Explain why_____. (Explain how_____.) • What would happen if _____? • What are the strengths and weaknesses of that assumption?
4. Questions that probe reasons and evidence The purpose of this question is to explore whether facts and observations support an assertion or conclusion.	• What would be an example that supports the evidence? • What are you assuming to be true when you say this is evidence? • What do you think causes _____? Why? • What evidence is there to support your conclusion? • Have you examined the evidence for any fallacies in logic?

Continues

**Six Types of Socratic Questions and
Examples of CTQs** (*Continued*)

5. Questions that probe viewpoints and perspectives The purpose of this question is to learn how things are viewed or judged and consider things not only in a relative perspective but also as a whole.	• What is a counterargument for _____? • What are the strengths and weaknesses of that viewpoint? • What are the similarities and differences between your point of view and someone else's point of view? • Compare _____ and _____ with regard to _____. • What is your perspective on why it happened?
6. Questions that probe implications and consequences The purpose of this question is to help understand the inferences or deductions and the end result if the inferred action is carried out.	• What are the consequences if that assumption turns out to be false? • What will happen if the trend continues? • Is there a more logical inference we might make in this situation? • How are you interpreting her behavior? Is there another possible interpretation? • Could you explain how you reached that conclusion? • Given all the facts, is that really the best possible conclusion?

When applying the example questions on the right-hand side of the above table, make them as specific as possible to the problem at hand. Make it clear which assumption or viewpoint you are challenging as is done in the following example about a new energy drink.

Concerns about a New Energy Drink

A new energy drink is on the market that combines vitamins with staying-alert power, while other energy drinks contain no vitamins. The company said the new drink had all the daily requirement of vitamins needed to stay healthy and feel energized. The drink's ability to keep people awake works especially well for college-age adults and pretty well for older adults. A study shows no harmful effects were observed in the vast majority of test subjects. While slightly more expensive than the other energy drinks, it is affordable to those who need it.

1. **Questions about the question/problem statement:**
   Why do we need to add to the cost by adding vitamins to the energy drink?
   Is there room in the market for another energy drink?
   How many of the test-market cases caused a harmful effect and what was the effect?

2. **Questions for clarification:**
   Here we see a number of ambiguous words or phrases:
   How are you defining "staying-alert" power?
   What "harmful effects" did the study look for?
   What does "feel energized" mean?
   What is a "vast majority"?
   What does "slightly more expensive" mean?
   What is "well affordable"?

3. **Questions that probe assumptions:**
   What would happen if consumers don't see a cost-effective advantage to the added vitamins?
   Will consumers believe that the new drink is safe because the "vast majority" suffered no harmful effects?
   Do consumers perceive that they need another source of vitamins?
   Can you explain why you think consumers will be willing to pay the cost differential for the new drink?

4. **Questions that probe reasons and evidence:**
   What marketing data suggest that the consumer will want vitamins in the drink?
   What evidence is there that the consumers are getting greater benefits from the new drink?

5. **Questions that probe viewpoints and perspectives:**
   What are the most positive and negative consequences of bringing the new energy drink to market?
   What are the similarities and differences with drinks currently on the market?
   What are the advantages and disadvantages of the product over products now on the market?

6. **Questions that probe implications and consequences:**
   Why is the drink not as effective in energizing older adults?
   Are there any dangers of taking vitamin supplements every morning?

Before moving on to critical thinking actions, let's look at the following reconstruction of a case history of a real-life example where Socratic questioning was used to uncover the real problem.

Dead Fish

Application of Critical Thinking Using Socratic Questioning

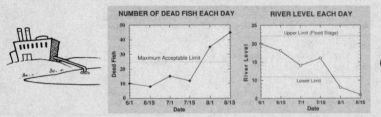

Chris Shannon is a waste treatment manager with eight years of experience with the company. One day the section head comes into Chris's office and says, "We need to design a new waste treatment plant to reduce the toxic waste stream flowing into the river by a factor of 10." Chris carries out a quick back-of-the-envelope calculation and realizes that the plant could cost several million dollars. Chris is really puzzled because the concentrations of toxic chemicals have always been significantly below governmental regulations and company health specifications that are even stricter than the recommendation of the Environmental Protection Agency. Has Chris been given a real problem or a perceived problem to solve?

Let's apply CTQs to this situation.

1. **Chris begins with a question about the question:**

   Q. Chris asks his supervisor, "Where did the problem originate?"
   A. His supervisor says it came from bad publicity in the newspapers.

2. **Chris knows that someone in the company must have read the newspaper for the company to be acting on the problem and asks a question for clarification:**

   Q. He asks his supervisor, "Who posed the problem in the first place?"
   A. The supervisor says, "Upper management."

3. **Chris thinks that the newspaper might not have included all the facts and asks a question that probes assumptions:**

   Q. Chris asks his supervisor, "Can you explain the reasoning management used to arrive at the problem statement?"
   A. The supervisor explains that fish are dying because of the low water level caused by an ongoing drought. Toxic chemicals become more concentrated—and hence more toxic—when the discharge is the same but the water level is lower.

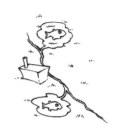

4. **Chris wants to know how strong the assumption is about the fish dying due to toxic chemicals and asks a question that probes reasons and evidence:**

   Q. Chris asks whether the concentration of chemicals in the river is approaching the LD50 level (meaning that 50% of the fish will die at this concentration).

   A. Chris is informed that the concentration in the river was not measured.

5. **If the LD50 level is not being reached, Chris thinks the assumption that the fish are dying due to a toxic level of chemicals is weak. Chris calls a biology professor at a state university and asks a question about viewpoints and perspectives:**

   Q. Chris asks the professor, "Is there is an alternative explanation as to why the fish are dying?"

   A. The professor explains that the low water levels and higher water temperatures make fish more susceptible to disease—perhaps fungi in this case.

6. **Chris wonders whether there are other locations in the area where fish are dead or sick, such as upstream of the plant or in surrounding lakes and rivers where the toxic chemicals are not present. Chris calls the state Department of Natural Resources (DNR) and asks a question that probes implications and consequences:**

   Q. Chris asks the DNR, "Have the fish upstream of the plant or in surrounding lakes and rivers where the toxic chemicals were not present been dying?"

   A. A government official at the DNR says dead fish have been found upstream of the plant and in nearby lakes.

7. **Chris nows knows that there is no way the toxic chemicals could diffuse upstream of the chemical plant or get into surrounding lakes. Chris asks a question that probes reasons and evidence:**

   Q. Chris asks the DNR, "Did the dead fish tested show any fungi or strange bacteria?"

   A. The DNR replies to say that the fish were infected with fungi in both the river and the lakes.

Epilogue

Chris's company was grateful that the real problem had been uncovered and that they did not go ahead and try to solve the perceived problem by building the multimillion-dollar plant. In regard to solving the problem about the fungi that are causing the fish to die, the company will leave that to the DNR.

Keep digging to learn the motivation (who, why) for issuing the instructions to solve the perceived problem.

Another component of critical thinking is the actions that one takes. Rubenfeld and Scheffer[4] list seven types of **critical thinking actions**, shown in the following table.

Types of Critical Thinking Actions	Examples of Critical Thinking Actions
1. Predicting: envisioning a plan and its consequences	I could imagine that happening if I . . . I anticipated … I was prepared for … I made provisions for … I envisioned the outcome to be … My prognosis was … I figured the probability of … I tried to go beyond the here and now …
2. Analyzing: separating or breaking a whole into parts to discover their nature, function, and relationships	I dissected the situation … I tried to reduce things to manageable units … I detailed a schematic picture of … I sorted things out … I looked for the parts … I looked at each piece individually …
3. Information seeking: searching for evidence, facts, or knowledge by identifying relevant sources and gathering objective, subjective, historical, and current data from those sources	I made sure I had all the pieces of the picture … I knew I needed to look up, or study … I wondered how I could find out … I asked myself if I knew the whole story … I kept searching for more data … I looked for evidence of … I needed to have all the facts …
4. Applying standards: judging according to established personal, professional, or social rules or criteria	I judged that according to … I compared this situation to what I knew to be the rule … I thought of/studied the policy for … I knew I had to … There are certain things you just have to account for … I thought of the bottom line that is always … I knew it was unethical to …

5. **Discriminating:** recognizing differences and similarities among things or situations and distinguishing carefully as to category or rank	I grouped things together … I put things in categories … I tried to consider what was the priority of … I stood back and tried to see how those things were related … I wondered if this was as important as … I thought of the discrepancies in the study … What I heard and what I saw were consistent/ inconsistent with… This situation was different from/the same as …
6. **Transforming knowledge:** changing or converting the condition, nature, form, or function of concepts among contexts	I wondered if that would fit in this situation … I took what I knew and asked myself if it would work … I improved on the basics by adding … At first I was puzzled; then I saw that there were similarities to … I figured if this was true then that would be too.
7. **Logical reasoning:** drawing inferences or conclusions that are supported in or justified by evidence	I deduced from the information that … I could trace my conclusion back to the data … My diagnosis was grounded in the evidence … I considered all the information and then inferred that … I could justify my conclusion by … I moved down a straight path from the initial data to the final conclusion … I had a strong argument for … My rationale for the conclusion was …

Let's now apply Rubenfeld and Scheffer's[4] seven critical thinking actions to expand on the case of the dead fish.

Critical Thinking Actions

 1. Predicting: envisioning a plan and its consequences

Chris **envisioned** that the proposed plant would cost millions of dollars and wanted to make sure that such an expenditure would solve the perceived problem.

Continues

2. Analyzing: separating or breaking a whole into parts to discover their nature, function, and relationships

Chris examined the available data presented by his supervisor and **sorted out** the relevant information and facts from perceptions to find that it was the newspaper data and not actual measurements of contamination that prompted the order of the new waste treatment plant.

3. Information seeking: searching for evidence, facts, or knowledge by identifying relevant sources and gathering objective, subjective, historical, and current data from those sources

Chris **contacted** the biology professor to learn possible causes for the dead fish problem.

4. Applying standards: judging according to established personal, professional, or social rules or criteria

Chris attempted to find out if the concentration of toxic chemicals in the river was above the **standard LD50**.

5. Discriminating: recognizing differences and similarities among things or situations and distinguishing carefully as to category or rank

Chris analyzed the fish kill data and **grouped them according** to **location**. Chris determined that the other locations in which fish are dying could not be affected by the plant's discharge. Chris questioned whether the fish in all the locations were dying from the same cause.

6. Transforming knowledge: changing or converting the condition, nature, form, or function of concepts among contexts

Chris recalled news items in the past where fish and other forms of water life have been harmed solely by natural causes and **wondered if that might apply to the current situation**. Chris contacted the state biologist. She informed him that fungi had indeed been found in several areas of water that were reporting high levels of fish dying and that this, coupled with the recent weather conditions, could be killing off the fish.

7. Logical reasoning: drawing inferences or conclusions that are supported in or justified by evidence

Chris **deduced** that it was possible that the fish are dying in the river owing to a fungal infection, rather than because of high levels of toxic chemicals.

Deeper Thinking

Don't close your mind just because you think you have found a good solution.[5] Although the first solution may appear to solve the problem, you must resist the temptation to blindly implement the solution. It is necessary to be aware of unintended consequences that can crop up: the hidden assumptions or additional alternatives that may present themselves before or during implementation of the solution. The following is a real-life example that illustrated the need to apply critical thinking actions and Socratic questions (CTQs) through the problem-solving process.

On the job, it is important to not get stuck in an infinite loop of constantly second-guessing your solutions to the point where no progress is ever made. However, as the case study shows, deeper thinking can save time and energy by bringing up potential pitfalls before they arise. Even when pressed for time, taking even just a few moments to dive into deeper thinking is almost always worth it.

Blind to the Cause

Pepsi-Cola developed an advertising campaign, "Take the Pepsi Challenge," that it hoped would increase Pepsi's market share at the expense of its bitter rival, Coca-Cola. As part of this campaign, Pepsi set up stations at various locations, such as shopping malls, where it conducted blind tastings of both Pepsi and Coke. As the individuals took the taste test, they were videotaped. The television commercials focused on videos where the tasters chose Pepsi. The management at Coca-Cola was very concerned about this advertisement, which showed Pepsi as the overwhelming favorite.

Find a solution to address Coke's concern about Pepsi's advertising campaign.

Let's develop a hypothetical critical thought process for how Coke could have responded to the Pepsi campaign. First let's look at the *critical thinking actions* Coke might have taken.

Information seeking: How many tasters chose Coke but were not shown on TV? Are there precedents for this type of consumer-comparison commercial? Has it affected market share of either product? Would the results be the same if

---

Continues

the tasters were required to drink at least eight ounces of both Pepsi and Coke before choosing?

Analyzing: Did Coke look for unstated or hidden assumptions? For example, was there an outside influence, such as the geographical part of the country where Pepsi is known to be popular; temperature (was it a hot day, was the Pepsi cold and the Coke warm?); time of day (just before or after lunch?); age of the tasters; presentation of the drinks by personnel at the stations?

Discriminating: What was it about the taste of the Pepsi that the tasters preferred? Was it sweeter? More carbonated?

Predicting: With increased showing of the Pepsi TV commercials, will Coke lose market share?

Now let's look at Coke's response to the advertising campaign. Coke quickly responded by coming out with a new product with a "better" taste. The preceived problem statement: **Make a new soft drink to challenge the perceived preferred taste of Pepsi.** *They dubbed this new product "New Coke" and put it on the market to replace the existing version of Coke.*

Did Coke apply Socratic questions, such as the following?

Question the solution of developing and marketing New Coke.

Which problem does creating New Coke solve? *(Question that explores viewpoints and perspectives)*

Will New Coke be able to negate Pepsi's advertisement campaign? *(Question that probes assumptions)*

Will New Coke increase or maintain Coke's market share? *(Question that probes reasons and evidence)*

What are the implications of creating New Coke in terms of cost, marketing, and acceptance? *(Question that probes implications and consequences)*

Did Coke carry out a *potential problem analysis* (see Chapter 8) of what could go wrong with their decision or probe the assumption that the change to New Coke will be for the better?

Were other solutions proposed for competing with Pepsi? *(Question for clarification)*

Examples of other solutions might be the following:

- Create a new advertising campaign by videotaping a tasting in a predetermined location and demographics where it is known that Coke will be preferred.
- Create a new advertising campaign by videotaping a tasting where the two drinks are not compared and the tasters are only asked what they like about Coke.

Coke's initial solution turned out to be one of the biggest product change mistakes in history. Customers tried New Coke but did not like it. The product was withdrawn from the market after 77 days and was replaced by the original Coca-Cola recipe, which was called "Classic Coke."

Subsequent studies showed that successful taste tests of Pepsi and New Coke versus Classic Coke did not suggest that people wanted an entire serving of the new formula(s). The increased sweetness of Pepsi and New Coke made them beat Classic Coke in taste tests when small quantities of the products were consumed. However, when drinking an entire serving, the preference switched to Classic Coke.

Challenge the Problem Statement

The real problem statement should have been "Develop a marketing strategy to regain Coke's market share."

SUMMARY

This chapter focuses on thinking skills that, if studied and practiced, will serve you well throughout your entire life, regardless of what discipline or job you choose.

- **Structured Critical Reasoning (SCR)**
  The sequence of the analysis is to identify:
  - Conclusions
  - Evidence
  - Assumptions
  - Strengths and weaknesses of each assumption
  - Fallacies in logic

- **The 11 Fallacies in Logic**
  - Ambiguous or vague words or phrases
  - Citing a questionable authority
  - Straw person
  - False dilemma
  - Red herring
  - Slippery slope

Continues

- – Appeal to popularity
- – The "perfect" solution
- – False, incomplete, or misleading facts or statements
- – Causal oversimplifications
- – Hasty generalizations

- **Six Types of Socratic Questions**
  - – Questions about the question/problem statement
  - – Questions for clarification
  - – Questions that probe assumptions
  - – Questions that probe reasons and evidence
  - – Questions about viewpoints and perspectives
  - – Questions that probe implications and consequences

- **Critical Thinking Actions**
  - – Predicting
  - – Analyzing
  - – Information seeking
  - – Applying standards
  - – Discriminating
  - – Transforming knowledge
  - – Logical reasoning

WEB-SITE MATERIAL (WWW.UMICH.EDU/~SPCS)

- **Learning Resources**
  Summary Notes
  Self-Tests
  1. Matching the Socratic Question Example to a Definition
  2. Identifying the Type of Socratic Question
  3. Matching Critical Thinking Actions

- **Professional Reference Shelf**
  1. Structured Critical Reasoning (SCR) Examples
     a) The Draft Once Again
     b) Downed Powerlines
     c) Continents in Motion
  2. Critical Thinking Questions
     a) Fires in Orange County
     b) A Real Toothache

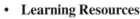

REFERENCES

1. Browne, M. Neil, and Stuart M. Keeley, *Asking the Right Questions: A Guide to Critical Thinking*, 10th ed., Pearson Education, Upper Saddle River, NJ, 2012.

2. Bornstein, Adam, "Do Eggs Cause Heart Disease?" www.livestrong.com/blog/do-eggs-cause-heart-disease/; Fung, Brian, www.theatlantic.com/health/archive/2012/08/study-eggs-are-nearly-as-bad-for-your-arteries-as-cigarettes/261091/; Spence, David J., David J. A. Jenkins, and Jean Davignon, "Egg Yolk Consumption and Carotid Plaque," *Atherosclerosis*, 2012, http://dx.doi.org/10.1016/j.atherosclerosis.2012.07.032; Ward, Elizabeth M., "Are Eggs the New Cigarettes?" http://blogs.webmd.com/food-and-nutrition/2012/08/are-eggs-the-new-cigarettes.html.

3. Paul, R. W., *Critical Thinking*, Foundation for Critical Thinking, Santa Rosa, CA, 1992.

4. Rubenfeld, M. G., and B. Scheffer, *Critical Thinking TACTICS for Nurses: Achieving the IOM Competencies*, Second Edition, Jones & Bartlett Publishers, Sudbury, MA, 2010.

5. Zyman, S., and A. A. Brott, *Renovate Before You Innovate*, Portfolio, a member of the Penguin Group, New York, 2005.

EXERCISES

3.1. Carry out a structured critical reasoning (SCR) analysis on the following situation. It has been recently shown that gun control works better in small towns or villages than in large cities. Small towns have more of a family atmosphere and people are more likely to help one another; as a result people never have a need to feel unsafe or need firearms. The mayor of Long Bridge, Michigan, population of 440, has said in this day and age no one needs a firearm because if trouble did arise, they have one of the best law enforcement officers, Sheriff Bradshaw, to take care of it.

3.2. Apply SCR to the following Editorial Opinion (EdOp).

People are flocking to buy electric cars, and they will soon take over the mass market. The electric car company Tesla's stock has increased 250 percent in the past six months, and electric car charging stations can now be seen in Whole Foods parking lots, as well as a few University of Michigan parking lots in Ann Arbor. Ryan Davis, a very successful businessperson and proud owner of a new Tesla, said that the Tesla is inexpensive, with model prices starting at $60,000, compared with the new BMW 750 Li, which he also considered and which sells for $100,000. He continued, "Soon, charging stations will be as common around the city as parking meters. Besides, people want to reduce their carbon footprint and dependence on foreign oil, and electric cars will help do that. I expect Tesla sales to increase by a factor of 3-4 over the next several years."

Sergey Furtaev/Shutterstock

3.3. Apply SCR to the synopsis of the editorial opinion article related to "Why We Don't Talk Anymore? Tracking Phone Call Lengths."

"'It is,' once mused an actor of London's east-end origins, 'good to talk'." Since mobile phones tipped into the mainstream in the late Nineties, we've had voice contact with everyone from loved ones to the local pizza delivery place a pocket's distance away. But, according to the CTIA, the trade group representing the U.S. wireless industry, the average length of our mobile phone calls has dropped drastically in the last six years. In 2006 the average call was 3.03 minutes long. By the end of 2011 they were down to 1.78 minutes. Why have we stopped talking (or, at least, paying to talk)?

The answer, at least according to a lengthy report in *The Wall Street Journal*, is—like so many things—attributed to Apple and its 2007 release of the iPhone, which allowed users to communicate via numerous non-call routes including voice-over-internet protocol, email and (the fee-free) iMessage. The other smartphones that followed Apple only furthered this troublesome development for the phone networks (who will counter it with increased unlimited-call packages).

It's not just smartphones, though. Since 2006, Facebook has gained as many users as there were on the entire Web at its inception in 2004. Twitter has no doubt eaten into the SMS market, too. Do we need to ring cousin Dave to see how his newborn is doing, when we can see pictures of the baby on Facebook and get instant updates from the delivery room? Possibly not, but we won't stop chatting. The launch of Sean Parker and Shawn Fanning's peer-to-peer chat network, Airtime, hopes to do to the phone industry what their Napster did to music. This could be great news, unless you own shares in AT&T."

Source: *The Independent* (a British newspaper), June 7, 2012.

3.4. Carry out an SCR analysis on the following memo.

Memorandum

To: Harley Davidson, Director of Housing

From: Natalie Dressed, Northend Residence Hall Manager

Subject: Northend Residence Hall Problems

The Northend residence hall was completed in the center of campus just in time for the start of the 2010 Fall term. After the first week of classes the resident dorm counselor told the supervisor of all university housing that the dorm was in chaos. The counselor complains it is so overcrowded that they never should have built so many dorm rooms on each floor. The students have to live in cramped quarters, often with up to three students living in a room. The cafeteria is a disaster as is clearly demonstrated by the long lines to get served and the almost shoulder-to-shoulder crowds at lunch hour. The cafeteria ran out of food on one of the five days during the first week of class after it opened. There was such poor planning that students were sitting on the floor with their food trays on their laps, a violation of the health code. Backpacks were piled up, blocking the walking spaces between the dining room tables, a violation of the fire code.

The counselor, busy with putting out other "first-week fires," has not visited the cafeteria in the morning or evening but imagines it is even worse then. However his friend John stopped by Friday night and said that it was not crowded at all, perhaps due to a football pep rally over by the Union, otherwise it would be mobbed the same as it is at lunchtime. There are just too many students in the new dorm. The university should remodel some of the three-bedroom suites to turn them into two-bedroom suites so there will be less crowding.

3.5. A survey of the students in the senior-year Engineering 405 class showed that more than 75% of the students either now play or have played a musical instrument for at least three years. Consequently, taking three years of music lessons sometime during the K–12 years will better prepare you for a career in engineering. Write CTQs you would use to challenge this conclusion. State the type of CTQ you are asking.

3.6. Choose one or more of the following SCR examples and make a list of CTQs you would ask to challenge the reasons and evidence in each example. For example, in "Truancy in U.K. Schools" the evidence is given that "eight million days of school were missed last year according to the Department of Children, Schools, and Families." A CTQ to challenge this would be a question probing the assumption that this is out of the ordinary: "Is this always the case?" A question probing reasons and evidence is "How many days of school were missed two years ago?"

1. "Fighting Fires?"
2. "Truancy in U.K. Schools"
3. "A Public Health Hazard—Eggs"

3.7. Review the perceived problem/real problem examples in Chapter 1. Make a list of CTQs you would have asked for one or more of the following:

1. "Better Printing Inks"
2. "Making Gasoline from Coal"
3. "A Picture Is Worth a Thousand Words"
4. "Dam the Torpedoes or Torpedo the Dam?"

3.8. You are an MI6 agent assigned to investigate the following case, "Spy Found Dead in a Bag."

A. Make a list of the critical thinking actions and questions that you would use in order to gather information about this case. In formulating your CTQs, state to whom you would ask each question.

B. Carry out an SCR analysis.

"[An] MI 6 officer was found dead stuffed inside a padlocked duffle bag at his central London flat. His flat was very secure as only 'vetted' people were admitted to building. The flat showed no signs of forced entry and everything seemed in place except a red female wig hanging on the back of a chair. They also found £20,000 worth of women's clothes in his closet. The MI 6 agent, a math prodigy who received his Ph.D. at age 21, was a code breaker who one of his peers described as extremely conscientious and the most scrupulous risk-assessor he had ever known. The dead agent was due to leave Central London to join the eavesdropping agency in Cheltenham as he had told his sister he had become unhappy with the London office culture of post-work drinks, competition and rat-race at the office. He said that they had dragged their feet on his transfer out of the office for several months until the London office spy chiefs agreed to the transfer a week ago.

Make-up and lipstick that were described as being in pristine condition were also found in his apartment along with the wig. His lifelong girlfriend insisted he was not a cross-dresser. His naked body was stuffed in a red North Face duffel bag and there appeared to be no signs of a struggle as his hands were folded across his body and his face was calm. The cause of death is uncertain because of the length of time between his death and the discovery of his body. DNA different from the victims was found on the padlock on the duffel bag. The agent was fit and muscular and an avid cyclist. One of the investigators suggested that a third party must have been involved.

The landlady reported that three years earlier she had heard him shouting in the early morning hours and got the key to his room and she and her husband went in to look and found him with both hands tied to the bed board with a knife on the table beside the bed. When asked, 'What the bloody hell are you doing?' he replied that he just wanted to see if he could get free. The husband said, 'We can't have this here,' and cut him free."

Sources: From *London Metro*, Morning Edition, April 24, 2012, page 5; *London Metro*, April 26, 2012, page 11; and *London Metro*, May 2, 2012, page 31.

3.9. "Mysterious Disease"

During spring 2011, an enterohemorrhagic E. coli (EHEC) infection spread throughout Europe. The bacteria infected thousands of people and several died from the disease. Everyone that was infected had been in Germany and had eaten vegetables in Germany.

While the infection struck visitors to Germany, imported cucumbers from Spain were blamed for the outbreak. These were later excluded as the source, and the false accusations cost the Spanish farmers around 280 million dollars. The food market grew skeptical toward European vegetables, and Russia totally stopped the import of vegetables from European countries during this time.

It was found that the probability of being infected was nine times higher among people who had eaten bean sprouts from Germany. This source was later excluded as well.

Although there was no definite proof that vegetables were causing the infection, a high-ranking official in the German government told farmers they had to destroy their harvest, losing hundreds of thousands of dollars each.

Which type of critical thinking action is each of the following?

A. Blaming bean sprouts as the cause because the probability of being infected was nine times higher among people who had eaten bean sprouts from Germany _____

B. Applying safety precautions from other food epidemics _____

C. Finding out what method the German scientists used to detect the bacteria _____

D. Finding out what bacteria caused the illness _____

E. Blaming German vegetables _____

F. Finding out what vegetables the infected people had eaten _____

3.10. Choose one of the following statements. Take a side for or against it, and prepare a one-page argument, keeping in mind that other readers would use the SCR method and Socratic questioning when examining your argument. Alternatively, for a class exercise, prepare a three-minute PowerPoint presentation describing your argument.

A. For higher education, online courses are a more beneficial option for students than universities.

B. Coke is better than Pepsi.

C. The legal drinking age should be 18.

D. Reality TV does more good than harm.

E. The United States should establish a colony on the moon.

F. Ultimate Frisbee should be an Olympic sport.

G. Autumn is the best season.

H. Single-sex schools are good for K–12 education.

    1. Turn in the SCR analysis for your chosen debate topic along with a number of Socratic questions that would challenge your presentation.

    2. For a class exercise, your instructor will distribute the SCR topics of the other class members. Prepare a critical thinking question for each of the topics chosen by the other members of the class.

3.11. Socratic Questions: Pick three of the issues listed in the debate issues in problem 3.10 and imagine you are going to see a presentation about one of them. Prepare three Socratic questions you would want to ask presenters of arguments for each topic.

Group Topic _____

Question 1. _____ (Category _____)

Question 2. _____ (Category _____)

Question 3. _____ (Category _____)

Group Topic _____

Question 1. _____ (Category _____)

Group Topic _____

Question 1. _____ (Category _____)

3.12. Match the question with the type of Socratic question:

Critical Thinking Questions

- How are ___ and ___ similar? Answer: (5) Questions about viewpoints and perspectives
- Why do you say that? _____
- What is the difference between ____ and ____? _____
- Compare ____ and ____ with regard to____ . _____
- What could we assume instead? _____
- What was the point of this question? _____
- What would be an example? _____
- What would be an alternative? _____

3.13. A student comes to the professor's office to say that her group did not get the team assignment finished. She says that one member of the group of four is not carrying his fair share of the load and is coming to meetings unprepared. She goes on to say that another group member is an effective team member but has missed about one-third of the group meetings. List as many CTQs as you can that the professor should ask the student. Identify the category for each question.

   1. _____
   2. _____
   3. _____
   4. _____

3.14. Match the following critical thinking action with the appropriate type of critical thinking action.

Critical Thinking Actions

I dissected the situation … Answer: (1) Predicting: envisioning a plan and its consequences.

I knew I had to compare … _____

I grouped things together … _____

I made sure I had all the pieces of the picture … _____

I deduced from the information that … _____

I could imagine that happening if I did … _____

Although this situation was somewhat different, I knew … _____

3.15. "Finding Out Where the Problem Came From: The Deepwater Horizon Oil Spill"

   A. Prepare six CTQs (one for each type) you would have asked during the drilling (or after the accident if you prefer) of the Deepwater Horizon accident described in the following text.

1   Deepwater Horizon, an ultra-deepwater offshore oil drilling rig, had been leased to BP
2   ever since its maiden voyage [in] 2001. BP used it for drilling in the Gulf of Mexico and
3   the rig was reported to cost as much as $600,000 per day.
4   BP, one of the largest companies in the world, has had a spotty reputation for safety.
5   Among other BP accidents is an explosion at a Texas refinery in 2005 where 15 workers
6   died, and in 2006 there was a major oil spill from a badly corroded BP pipeline in Alaska.
7   In April 2010, the BP team onboard the Deepwater Horizon worked at the Macondo
8   well. According to their plan, they decided to skip the usual cement evaluation if the ce-
9   menting went smoothly. Generally, the completion rig would perform this test when it re-
10 opened the well to produce the oil the exploratory drilling had discovered. The decision
11 was made to send the cement team home at 11:00 a.m. on the 20th of April, thus saving
12 time and the $128,000 fee. BP Wells Team Leader John Guide noted, "Everyone involved
13 with the job on the rig site was completely satisfied with the [cementing] job."
14   The rig crew began the negative-pressure test. After relieving pressure from the
15 well, the crew would close it off to check whether the pressure within the drill pipe would
16 remain steady. But the pressure repeatedly built back up. As the crew conducted the test,
17 the drill shack grew crowded. The night crew began arriving to relieve the day shift, and
18 some VIP's from a management visibility tour came in as part of their guided tour around
19 the platform. There seemed to be a problem but "tool pusher" Jason Anderson insisted
20 that senior tool pusher Randy Ezel should go and eat with the dignitaries before going off
21 his shift, being sure Anderson would call him if there was a problem. Tool pusher Wyman
22 Wheeler was convinced that something wasn't right, but had to go off his shift and leave
23 the situation in the hands of the night shift.
24   Later the same evening assistant driller Steve Curtis called Senior Toolpusher Randy
25 Ezel who had left his day pass. "We have a situation…. The well is blown out…. We have
26 mud going to the crown." Ezel was horrified. "Do y'all have it shut in?" Curtis: "Jason is
27 shutting it in now… Randy, we need your help." Ezell: "Steve, I'll be—I'll be right there."
28   At approximately 9:45 p.m. on April 20, 2010, methane gas from the well, under
29 high pressure, shot all the way up and out of the drill column, expanded onto the platform.
30 The gas reached the engine room and a spark ignited the gas and the disaster ensued. Fire
31 engulfed the platform and the workers left for the lifeboats. Eleven of the workers on
32 Deepwater Horizon were never found.
33   The emergency disconnect switch (EDS), which was supposed to unlatch the blow-
34 out preventer (BOP) and shut the well, was turned on. Unfortunately the BOP did not seal
35 the well. A few brave men stayed at the burning platform to manually try to unlatch the
36 BOP, but without succeeding.
37   When the leak was finally stopped it had released about 4.9 million barrels of crude
38 oil. The total cost is still unknown but BP has established a trust fund of $20 billion to
39 cover expenses.

Sources: www.bp.com/sectiongenericarticle800.do?categoryId=9036575&contentId=7067541
www.epa.gov/BPSpill/
www.oilspillcommission.gov/sites/default/files/documents/
DEEPWATER_ReporttothePresident_FINAL.pdf

In (1) through (6) below identify the line number and then ask a specific question using the type of Socratic question identified.

    1. Question for clarification, line number: _____

    2. Question about the question, line number: _____

    3. Question that probes assumptions, line number: _____

    4. Question that probes reasons and evidence, line number: _____

    5. Question about view points and perspectives, line number: _____

    6. Question that probes implications and consequences, line number: _____

  B. Write a critical thinking question you would ask at line ____ (your choice).

  C. What critical thinking action was taken at line ____ (your choice)?

  D. What critical thinking action would you have taken at line ____ (your choice)?

3.16. Create an example of (a) three, (b) six, or (c) all of the 11 fallacies similar to the ones shown in the table "Eleven Fallacies in Logic to Look For."

3.17. Recall the "Blind to the Cause" example about Coca-Cola's failed attempt to respond to the Pepsi challenge by introducing New Coke.

  A. Suggest CTQs and critical thinking actions that Coke might have applied before making the decision to discontinue New Coke after only 77 days on the market. *Example: Why is New Coke not preferred when we tried to make it similar to Pepsi?*

  B. Brainstorm solutions that Coke could have done to not discontinue New Coke. *Example: Have Coke do its own testing with both drinks at the same temperature but either extremely cold or at a lukewarm temperature to see if the flavors vary with temperature.*

3.18. It is possible to use critical thinking actions and CTQs to resolve fallacies in logic. For example, in "Draft Once Again" on the Professional Reference Shelf on the Web site, we found Fallacy #9, "False, incomplete, or misleading facts or statements," because no budget information was provided to support the authors' supposed cost savings of the government using cheap labor to save money. A critical thinking action of "information seeking" could be used such as *Looking for more data on the possibility of cost savings*" or a question for clarification such as "Could the government's cost savings by using cheap labor *be explained more*?"

Describe which critical thinking action or critical thinking question you would use to resolve each of the fallacies found in the SCR examples.

  A. "Fighting Fires?"

  B. "Truancy in U.K. Schools"

  C. "A Public Health Hazard—Eggs"

3.19. Identify the type of fallacy in each of the situations below.

  A. Would you like red or white wine with your meal?

B. If you continue playing these video games all the time, your grades will plummet, you won't be able to get into college, and your health will suffer from a lack of fresh air and exercise.

C. If we just raise taxes on the rich, the debt problem will be solved.

D. I didn't like the first song on the album so I didn't even bother listening to the rest.

E. Despite complaints that the car manufacturer has received recently regarding their new model SUV's poorly designed interior, the company's vehicles have an outstanding safety record.

F. The restaurant is far away and we have a deadline coming up soon, but I still think we have plenty of time to eat there later.

G. That Band-Aid is only going to stop the bleeding temporarily: don't even bother putting it on.

H. I know that Weight Watchers is an effective diet for many people because of its balanced approach, but I don't have the willpower to give up all of my favorite foods so I don't think I'll join.

I. I got 6/10 on my quiz, so I excitedly told my mom I only got four wrong and she was so happy.

J. I found it on seven Internet sites, so it has to be true.

K. My mother always used to say, "Don't sit so close to the television, you'll hurt your eyes."

L. Because everyone thinks we should get paid more, management should just get moving and get us all the raise.

3.20. Carry out the interactive exercises in the Summary Notes for Chapter 3 on the Web site.

FURTHER READING

Browne, M. Neil, and Stuart M. Keeley. *Asking the Right Questions: A Guide to Critical Thinking*, 10th ed. Pearson, Upper Saddle River, NJ, 2012.

Rubenstein, M. Gaie, and Barbara K. Scheffer. *Critical Thinking in Nursing: An Interactive Approach*, 2nd ed. Lippincott, Philadelphia, PA, 1999.

Paul, Richard W., and Linda Elder. *The Miniature Guide to Critical Thinking Concepts and Tools*. Foundation for Critical Thinking, Santa Rosa, CA, 2009.

Paul, Richard W., and Linda Elder. *The Thinkers Guide to the Art of Critical Thinking*, 2006.

4   FIRST STEPS

The mere formulation of a problem is far more essential than its solution, which may be merely a matter of mathematical or experimental skill. To raise new questions, new possibilities, to regard old problems from a new angle requires creative imagination and marks real advances in science.

—Albert Einstein

The First Four Steps

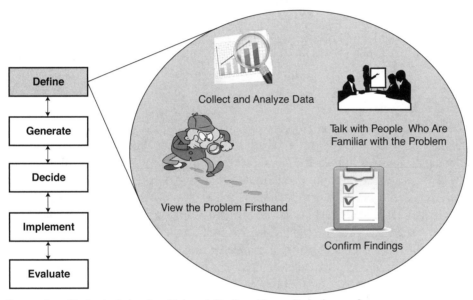

(Images from Shutterstock: iconico, Aleksandr Bryliaev, Nevena Radonja, gnurf)

Often, one of the most difficult aspects of problem solving is understanding and defining the real problem (sometimes referred to as the underlying or root problem). In Chapter 1, we presented a number of real-world examples of incorrectly defined problem statements that demonstrated how even competent, conscientious people can define the wrong problem and waste considerable time and money looking in the wrong direction for a solution. The first task in defining the real problem is to properly gather information about the problem.

THE FIRST FOUR STEPS

The first four steps to understand and define the real problem focus on gathering information.

1. Collect and analyze information and data.
2. Talk with people who are familiar with the problem.
3. If possible, view the problem firsthand.
4. Confirm all findings and continue to gather information.

> *"Start with an open mind."*
>
> *"Don't jump to conclusions."*
>
> *"Look at the big picture."*
>
> *"Review the obvious."*

Step 1: Collect and Analyze Information and Data

Learn as much as you can about the problem. Write down or list everything you can think of to describe the problem. Do an Internet search on all aspects of the problem. Until the problem is well defined, anything might be important. Determine which information is missing and which information is extraneous.

Visualize Information

The information should be properly organized, analyzed, and presented so that it can serve as the basis for subsequent decision making. Make a simple sketch or drawing of the situation. Drawings, sketches, graphs of data, and other illustrations can all be excellent communication tools when used correctly. Analyze the data to show trends, errors, and other meaningful information. Display numerical or quantitative data graphically. Tables can be difficult to interpret and are sometimes misleading (see "Presentation of Data" in Appendix 3 on the Web site). Graphing, by comparison, is an excellent way to organize and analyze large amounts of data. The business research of a private car company provides an interesting example of the use of graphical data to solve problems.

Everyone's Private Driver

A private car company was looking to improve business by reducing the downtime of its drivers by connecting empty private cars to people who want rides. The drivers could make extra money rather than being idle and the passengers get private car service on demand. The company wanted to answer the question, "Where do people want rides in the city?"

The project team began with their own ride history data for their city so they would know where to send idle drivers for the best odds to find customers. They examined their 40 most common pick-up spots in table form but couldn't really make any conclusions.

Street Addresses of the Most Common Pick-up Locations			
104 Maple	8 Mercer	755 Colomy	74 Euclid
556 State	65 George	3673 Perry	56 Rutgers
73 Shyne	466 Durante	745 5th	65 George
45 5th	1233 Perry	754 Harl	7 Grady
832 Evans	25 Nickols	88 9th	74 Kane
7765 Markley	44 Rivington	74 Highland	83 Naper
2343 Ann	23 Charles	99 Almond	890 Madison
56 Rutgers	655 4th	88 Alma	23 Nickols
899 Jane	843 Madison	6443 Burns	34 Broadway
544 Ludlow	6443 Broone	413 Burley	466 Durante

They then moved the data to graphic form by plotting all of the ride locations on a map of the city. As you can see from the graphic, it is much easier to get a sense of where demand for rides is high with the plotted data than a list of ride data.

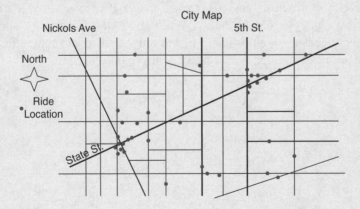

They company began deploying its drivers to the high-density areas, like the intersections of Nickols and 5th with State Street, and business increased immediately.

Order of Magnitude Calculation (Fermi Problems)

When collecting and analyzing data one would like to know what data are needed to calculate the final answer. Then, even before collecting data, one can use a Fermi analysis to obtain an order of magnitude estimate of how large the answer might be. This Fermi number can be used to validate your final answer, by checking that your final answer is within an order of magnitude of what you were expecting. Enrico Fermi was an Italian physicist who received the Nobel Prize for his work on nuclear processes. Fermi was famous for his "back-of-the-envelope order of magnitude calculation," which enabled him to obtain an estimate of the answer to a problem on as little space as the back of an envelope by applying logic and making reasonable assumptions for each number in the calculation of the answer. He used a process to set bounds on the answer by saying it is probably larger than one number and smaller than another, and he arrived at an answer that was within a **factor of 10.** When carrying out a Fermi calculation you should ask the following questions:

- What do you know is relevant?
- What assumptions can you make?
- How plausible are your assumptions?
- Is your chain of reasoning accurate?
- Can you do the problem another way and see if the result is the same?
- In your answer, did you spell out your assumptions, reasoning, solution, and checking procedure clearly?

Fermi calculations are useful when you need to get a reasonable estimate rather than a specific answer. It is important to be cognizant of poor assumptions and fallacies in logic as you estimate numbers, but in this case, the idea is to move forward quickly with a best guess to get your answer. If an exact answer is needed, a more thorough analysis can be done. For example, the National Association of Piano Tuners needs a rough estimate of the number of piano tuners in Chicago, rather than the exact number. A Fermi calculation can be used.

Fermi Example: Piano Tuners in Chicago

The National Association of Piano Tuners (NAPT) is launching a new training series, with its first training session taking place in Chicago. This is the first time such an event will take place, and the NAPT wants to contact all the piano tuners in Chicago to invite them to the training session where they will learn a new technique of piano tuning. The NAPT wants to get a rough idea of about how many piano tuners there are in Chicago so they know how much to budget for mailings and what size room to reserve. The event organizers use a Fermi calculation to get a number, which is outlined in the following.

The city of Chicago has around four million inhabitants. If we estimate that the average number of people living per household is four, then the total number of households in Chicago is around one million. Assuming one-fifth of the households own a piano, then there are 200,000 pianos in Chicago. If an additional assumption is made that a piano needs to be tuned once each year the total of tunes per year would be 200,000 in Chicago. Tuning a piano takes four to five hours, which is two per day. Furthermore, the piano tuners work approximately 250 days per year and will perform 500 tunes per year, since they tune two pianos each day. That brings us to the final calculation of the Fermi Problem calculation of the number of piano tuners in Chicago: 200,000 tunes per year divided by 500 tunes per year for each piano tuner gives us 400 tuners.

Population of Chicago: 4,000,000

Number of people per household: 4

$$\frac{4,000,000 \ people}{4 \ people \ per \ household} = 1,000,000 \ households$$

Continues

Number of households: 1,000,000

Fraction of households with pianos: 1/5

$$1{,}000{,}000 \; households \times \frac{1}{5} \; households \; own \; a \; piano = 200{,}000 \; pianos \; in \; Chicago$$

Households with pianos: 200,000

Number of tunes each piano needs per year: 1

$$1\frac{tune}{year} \times 200{,}000 \; pianos = 200{,}000 \frac{tunes}{year}$$

Tunes needed by pianos per year: 200,000

Number of days tuners work per year: 250

Number of tunes each tuner does per day: 2

$$250\frac{workdays}{year} \times 2\frac{tunes}{day} = 500\frac{tunes}{year}$$

Number of tunes by each tuner per year: 500

Number of tuners required for 200,000 tunes per year: 400

$$200{,}000\frac{tunes}{year} \times \frac{1 \; tuner}{500\frac{tunes}{year}} = 400 \; Tuners$$

Cautions: As you go along in your calculations, always question the strength of your assumptions and fallacies in logic. For example, "Does one out of every five households in Chicago really own a piano?" It's not all the households and it's not one out of 30. It's reasonable to believe that one out of five is pretty close, so we move forward.

The NAPT knew that the Fermi calculation was probably a bit lower than the actual value, because the calculations did not include churches and schools (see Exercise 4.8), so it planned the event accordingly.

Epilogue: The NAPT had 426 piano tuners for its training session in Chicago. The Fermi calculation of 400 allowed the NAPT to budget for mailings and reserve a room months ahead of time, without knowing the exact number.

Source: http://mathforum.org/workshops/sum96/interdisc/sheila2.html.

Step 2: Talk with People Who Are Familiar with the Problem

Find out who knows about the problem and interview them. Ask penetrating questions using six types of Socratic questions and by doing the following.

- Look past the obvious.
- Ask for clarification when you do not understand something.
- Ask when the problem first occurred.
- Ask how the problem occurred.
- Challenge the basic premise of any explanations given as to the cause of the problem. Ask for *reasons and evidence* to support their premise.
- Ask where the problem is located and where everything is okay.
- Ask what was observed and whether it had been observed previously.
- Ask who else knows about the problem.
- Probe the answers you receive with follow-up questions.
- Listen for what is **not** being said as well as what is being said.

Depending on the particular response you receive, additional probing with follow-up questions may prove particularly helpful. Our experience shows that seemingly naive questions (often perceived as "dumb" questions) can produce profound results by challenging established thinking patterns. This act of challenging must be an ongoing process.

You should also describe the problem to other people. Verbalizing the problem to someone else helps clarify in your own mind just what it is you are trying to do. Try to find out who the experts in the field are—and then talk to them.[1]

When interviewing people to collect data and information, try to use as many of the types of critical thinking questions as you can (see Chapter 3, "Skills Necessary for Effective Problem Solving"). The following example depicts a situation where Socratic questioning of an individual familiar with the problem is used to find and uncover the real problem.

Ask insightful questions.

The most important thing in communication is to hear what is NOT being said.
—Peter Drucker

The Anti-Crime Commissioner

A commissioner ran for office and won on the platform that he would clean up the streets and reduce crime throughout the city. He increased police patrol in high-crime areas and instituted sweeping changes in the judicial system to increase sentence durations; despite these changes, crime rates rose. The commissioner hired a private consulting firm and charged them with the task of solving the following perceived problem: *Find out why more crimes are being committed.*

The consulting firm interviewed the commissioner and some of his officers and asked the following series of Socratic questions to get to the answer.

Questions about the question

Do all crimes get reported in the crime-rate data?

The commissioner says that crimes get reported only when the person thinks the police can help them.

Questions for clarification

Is a higher "reported" crime rate necessarily a bad thing?

The commissioner realizes that a higher reported crime rate does not necessarily mean that more crimes are being committed. Instead, the people are more likely to report committed crimes due to increased confidence in the police department since the addition of more police patrols.

Questions that probe assumptions

Why did the commissioner assume that increased patrols would decrease the crime rate?

The commissioner says that with more police on the streets, the response to a reported crime would be more rapid, and criminals would be more likely to get caught and therefore less likely to commit crimes.

Questions that probe reasons and evidence

What is the commissioner's evidence that there is an increase in the crime rate?

The commissioner says the crime rate is the number of crimes reported in his city during a given time period and this number has increased.

Questions that probe viewpoints and perspectives

Do criminals ever think that with more police they would be more likely to be caught?

The commissioner says that most criminals report that they never thought that they would be caught.

Questions that probe implications and consequences

Have increased patrols and the new changes in the judicial system influenced the public's view of the police?

A town poll shows that the public is now much more confident in the ability of the police force.

Conclusion: The instruction given by the commissioner was to solve the perceived problem: Find out why the number of crimes committed has increased. The real problem was rather to find *why the reported crime rate had increased*. When the firm researched this problem, they found that people had more confidence in the police department so they reported more of the crimes that were committed.

When different individuals interviewed give different interpretations of a situation, it is imperative to be aware that their individual biases can obscure or mask the actual facts. Separate the answers into those that are fact, opinionated fact, and opinion.

Fact	
Facts usually come from measurements, reports, tables, figures, firsthand observations, and other data.	"On August 16th, 45 fish were found dead in the river."
Opinionated Fact	
Opinionated facts use phrases to put a bias or spin on the factual data, in an effort to persuade others to adopt a particular point of view.	The opinionated phrase may denote the significance of the fact ("*only* 45 fish died"), attach value to the fact ("*surprisingly* 45 fish died"), suggest generalization ("*all* 45 fish were killed by…"), or advocate acceptance of the fact ("*obviously*, 45 dead fish is a serious problem").
Opinion	
Opinion is based on years of experience and can be quite useful in analyzing the problem. It can also be used to serve a person's self-interests.	"I think you need to look for another reason other than the toxic waste to explain the 45 dead fish."

Go Talk to George

When equipment malfunctions, it is a must to talk to the operators because they know the "personality" of the equipment better than anyone. Most organizations have employees who have been around a long time and have a great deal of experience, as illustrated in the following example.

Remember the printing machines for the redesigned currency discussed in Chapter 1? The solution that the company adopted was to replace the printing machines rather than develop different inks. Let's consider a similar situation in which, immediately upon replacement, one of the printing machines began to malfunction. We list, in order, three people to whom we most likely would want to talk.

1. The technician who monitors the printing machine
2. The manufacturer's representative who sold you the printing machine
3. George

Who's George? Every organization has a George. George is the individual who has both years of experience to draw upon and "street smarts." George is an excellent problem solver who always seems to approach the problem from a different viewpoint—one that hasn't been thought of by anyone else. Be sure to tap into this rich source of knowledge when you face a problem. Individuals such as George can often provide a unique perspective on the situation. Strive to become the "George" of your organization.

Search out colleagues who may have useful information and pertinent ideas. Have them play the "What if …?" game with you: "*What if* the printing machine only printed five bad bills per day out of thousands; would we need to replace it?" or "*What if* the one malfunctioning printing machine was eliminated?" Also have them play the devil's advocate and deliberately challenge your own ideas. This technique stimulates creative interactions. Non-experts are also a rich source of creative solutions, as evidenced by the following example.

Seeking Advice

Joel Weldon, in his tape "Jet Pilots Don't Use Rearview Mirrors," described a problem encountered by a major hotel when it opened several decades ago. The elevator capacity was inadequate for the number of guests, causing a backup on each floor and in the elevator area of the lobby. Even with the installation of mirrors by the elevators, the guests and hotel management found the wait time unacceptable. The manager and assistant manager were lamenting the problem in the lobby one day and were brainstorming about how to increase the elevator capacity. Unfortunately, there were too many people for adding mirrors to work like in the hotel in Chapter 1. Adding additional elevator shafts would require removal of a number of rooms and a significant loss of income.

The doorman, whose name happened to be George, overheard their conversation and casually mentioned that it was too bad they couldn't just add an elevator on the outside of the building, so as not to disturb things inside. A great idea! It occurred to the doorman because he was outside the building much of the time, and that was his frame of reference. Notice, however, that the doorman's creativity alone was not enough to solve the problem. Knowledge of design techniques was necessary to implement his original idea. A new outside elevator was born, and the rest is history. External elevators have since become quite popular in major hotels. Information, good ideas, and different perspectives on the problem can come from all levels of the organization.

Source: *Chemtech*, 13, 9, p. 517, 1983.

Step 3: View the Problem Firsthand

While it is important to talk to people as a way to understand the problem, you should not rely solely on their interpretations of the situation and problem. If possible, inspect the problem yourself.

Memo Angeles/Shutterstock

> *"You can see a lot just by looking."*
> —Yogi Berra

Viewing the Problem Firsthand

In 2012, for the first time in history, a majority of the people in the world live in cities and it is estimated that as much as 70% of the world's population will live in cities in 2050. All cities are not created equally, however, and urban planners have been researching how to foster creativity and economic growth in metropolises. Initial research showed a direct correlation between population density and innovation, as measured by the number of successful start-up companies, among other metrics. Researchers rationalized that the more people there are in a given area, the more casual interactions there are between diverse ideas, which often leads to innovation and economic growth.

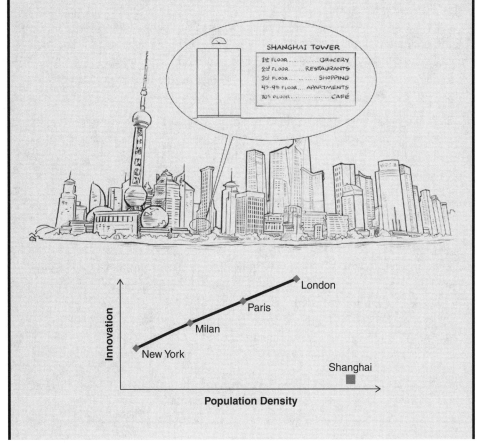

Continues

Cities such as New York, London, Paris, and Milan all fit well with the model of population density equating with innovation, but the urban planners were confounded by Shanghai. With nearly 125,000 people per square mile in some areas, Shanghai should rank highly in innovation according to the population density model; however, this is not the case. Planners went over and over all their maps, data, and criteria and could not figure out why it did not fit; New York, London, Paris, and Milan, all with smaller densities, outpace Shanghai in innovation. The team of urban planners working on the model thought that maybe their survey data were insufficient because Shanghai is such a high-density area. However, before they spent millions of dollars on another more extensive research project, they decided to view the problem firsthand on their own and flew to Shanghai.

The team toured the city extensively and could not see any reason why Shanghai lagged behind in innovation compared to the other cities. On the final day of the trip, the team members were going to have dinner, but it was raining heavily so they asked the hotel reception where they could eat close by. The receptionist said, "Well, you don't need to leave the building, you can eat at one of the seven different restaurants in the mall in the basement, the café on the top floor, or visit the grocery store on the ground level." The team was amazed and asked if most buildings had similar amenities. The receptionist said it was very common for buildings in the area to contain offices, residences, shopping centers, and more.

Shanghai has a high population density, but in order to accommodate all the people in such a small area, lives were compressed into even smaller areas with buildings functioning as self-sufficient suburbs. This reduction in required day-to-day mobility decreased the amount of casual interactions between people that would have led to innovation.

The urban planners followed this idea and found that the most innovative parts of other cities were not the skyscraper-packed streets of high-density population areas, but rather the slightly less population-dense neighborhoods and districts. They mixed multiple purposes, such as combining residential and business, pushing people to move around the city and come into contact with a larger number of ideas than if they simply stayed in a self-sufficient skyscraper. Since the team viewed the problem, in this case their outlying data point, they were able to revise their model to better plan future cities to limit the overcrowding that suppresses innovation and creativity.

Source: http://online.wsj.com/article/SB10000872396390443477104577551133804551396.html?KEYWORDS=for+creative+cities+the+sky+has+its+limit.

Step 4: Confirm All Key Findings

Verify the information you collected. Check and cross-reference data, facts, and figures. Search for biases or misrepresentation of facts. Confirm all important pieces of information, and spot-check others. Distinguish between fact and opinion. Challenge assumptions and assertions.

FreshPaint/Shutterstock

Confirm All Allegations

The authors of this book were involved in a consulting project for a pulp and paper company we will call Boxright. Several years ago, Boxright had installed a new process for recovering and recycling its chemicals that were used in the paper-making process. Two years after the installation, the process had yet to operate correctly. Tempers flared and accusations flew back and forth between Boxright and Courtland Construction, the supplier of the recycling equipment. Courtland claimed the problem was that Boxright did not know how to operate the process correctly; Boxright contended that Courtland's equipment was designed incorrectly. Boxright finally decided to sue Courtland for breach of equipment performance.

Much data and information were presented by both sides to support their arguments. Courtland presented data and information from an article in the engineering literature that the company claimed proved Boxright was not operating the process correctly. At this point it looked like Courtland had cooked Boxright's goose by presenting such data. Before conceding the case, however, Boxright needed to confirm this claim. When we analyzed this key information in detail, we found in the last few pages of the article a statement that the data would not be expected to apply to industrial-size equipment or processes. Had the allegations been accepted as presented and not confirmed, the suit would have been likely lost in favor of Courtland. Instead, when the article's key information was presented, the lawsuit was settled in favor of the pulp and paper company, Boxright.

Throughout the problem-solving process, you should continue to gather as much information as possible by reading texts and literature related to the problem to learn about both the underlying fundamental principles and any peripheral concepts. Literature searches are particularly helpful in this regard. Perhaps a closely related problem has already been solved. George Quarderer of Dow Chemical Company appropriately describes the idea of reinventing the wheel by his statement, "Four to six weeks in the laboratory can save you an hour in the library." The message is clear: Doing a bit of research into the background of the problem may save you hours of time and effort.

The information you gather in the first four steps will be of great help to you as you "define the real problem," which is the topic of the next chapter.

SUMMARY

In this chapter we discussed methods for obtaining the information needed to define the problem. Experienced problem solvers use four steps to attack problems:

1. **Collect and analyze information and data.**
   a. Visualize the data
   b. Use Fermi estimates

2. **Talk with people who are familiar with the problem.**
   a. Use Socratic questions
   b. Talk to George

3. **If possible, view the problem firsthand.**

4. **Confirm all findings and continue to gather information.**

WEB-SITE MATERIAL (WWW.UMICH.EDU/~SCPS)

- **Learning Resources**
  Summary Notes

- **Professional Reference Shelf**
  1. Viewing the Problem Firsthand
     – The Off-Spec Plastic
     – A Shocking Installation

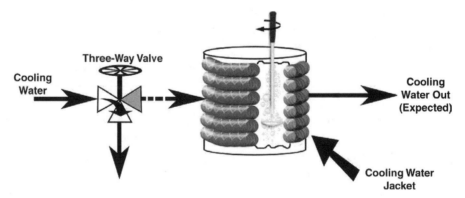

  2. Visualizing Information
     – The Case of the Dead Fish

REFERENCE

1. Browne, M. Neil, and Stuart M. Keeley, *Asking the Right Questions: A Guide to Critical Thinking*, Pearson Education, 2010.

EXERCISES

4.1. List the three most important things and the three most interesting things you learned in this chapter.

4.2. Describe a person from your past who has the characteristics of "George."

4.3. Describe a situation from your past in which you encountered a problem or situation and needed to collect information. Which of the four steps did you use?

4.4. Search the Internet to learn more about the Kansas City Hyatt Regency disaster discussed in the Professional Reference Shelf in Chapter 1. Do the same for "Spy Found Dead in a Bag" in the Chapter 3 exercises.

4.5. Write out a generic procedure you will use to gather information about a project or problem.

Fermi Problems

4.6. How many square meters of pizza were eaten by an undergraduate student body population of 20,000 during the fall term of 2015?

4.7. If we could convert all the weight that dieters lose in the United States every year to energy, how many gallons of gasoline would be its equivalent?

4.8. Refine the calculated number of piano tuners in Chicago by including pianos in churches and schools.

4.9. How many exams does a high school geometry teacher give throughout a career?

4.10. How much should you charge to wash all the windows in all the residential homes in Seattle?

4.11. How many vacuum cleaners are made a year?

4.12. Goodcase, a suitcase manufacturer, is launching a new extra large (XL) suitcase, in addition to its current small, medium, and large models. You are a plant engineer at Goodcase, in a meeting where the details of the XL are being rolled out. The additional raw materials for the XL have been ordered and are scheduled to arrive on time. The luggage makers have been trained to make the new model, and the new schedule with an extra shift has been created for the production of 50 XL suitcases per day, seven days a week. Additional trucks to ship the new suitcases are also scheduled to pick up suitcases every Sunday evening.

You are concerned that the rollout team is planning on storing suitcases in the extra supply room, dimensions of 10 ft × 10 ft × 30 ft, while they are waiting to be shipped. Suitcases come in boxes 3 ft by 2 ft by 1.5 ft.

Carry out a Fermi calculation to see if space will be an issue.

5 PROBLEM DEFINITION

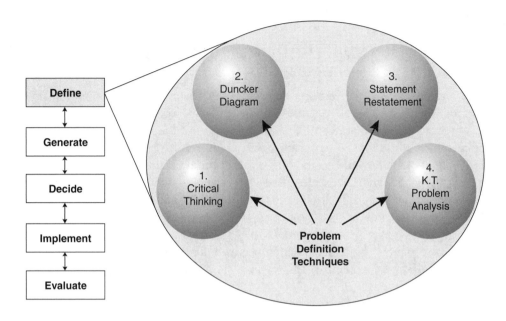

In Chapter 1, we described situations in which millions of dollars were wasted when individuals defined a *perceived problem* instead of the *real problem*. In this chapter, we address the first part of the problem-solving heuristic—problem definition. We present four techniques that will greatly enhance your chances of defining the real problem when combined with the critical thinking skills and information-gathering actions learned in the previous chapter.

The authors of this book conducted a study of the problem-solving methods of experienced engineers in industry. The study revealed some common threads that run through their problem definition techniques. We have classified these common threads into four techniques to help you understand and define the real problem instead of being sidetracked by a perceived problem.

DEFINING THE REAL PROBLEM: FOUR TECHNIQUES

The four steps discussed in Chapter 4 are all used to gather information about the problem. This information lays the groundwork that allows us to use the four problem definition techniques to help define the real problem instead of the perceived problem. In Chapter 3 we discussed one of these techniques, critical thinking.

In Chapter 5 we add three additional techniques to define the real problem. We will start with one that has been well received when we have presented it in industrial short courses over the last several years, the Duncker diagram.

TECHNIQUE 1: USING THE DUNCKER DIAGRAM TO DEFINE THE REAL PROBLEM

The unique feature of the Duncker diagram is that it contains two major pathways (general solutions) to go from the present state (the problem statement) to the desired state (an acceptable problem solution). Part of the Duncker diagram technique involves describing the **present state** (where you are) and then describing **the desired state** (where you want to go). The descriptions of these states are reworked until all concerns and needs identified in the present state are addressed in the desired state.[1] However, the desired state should not contain solutions to problems that are not in the present state. Sometimes the statements appear to match but they don't get to the heart of the problem or they allow for many solution alternatives. You must recognize when this situation occurs and continue to rework the present state/desired state statements. This process is called *cleaning up the problem statement.*

Cleaning up the problem statement.
Luis Santos/Shutterstock

When determining the desired state, avoid using ambiguous and vague words or phrases like "best," "minimal," "cheapest," "within a reasonable time," or "most efficient," because these words mean different things to different people. Be quantitative where possible. For example, say "The children's playground needs to be completed by July 1st, at a cost less than $50,000" rather than "The playground should be completed in a reasonable time at minimal cost."

There are two sides to a Duncker diagram.[1]

Side 1. General solutions on the left side of the diagram show us how to move from the present state to *achieve the desired state*.

Side 2. General solutions on the right side of the diagram show us how to *make it okay not to achieve the desired state*. (This idea may seem a bit contradictory— but it will be clear in a moment.)

There are two steps involved in each pathway: functional solutions that describe *what you need to do* and then specific solutions that describe *how to do it*. Functional solutions tell us what we could do to move from the present state to either *achieve* or *not achieve* the desired state. After a number of functional solutions are developed, we then generate a number of specific solutions that guide us in implementing *each* of the functional solutions. Specific solutions tell us how to implement the functional solutions.

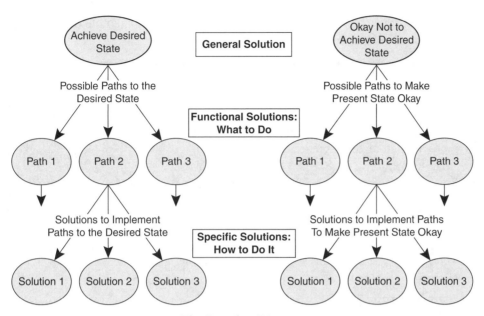

The Duncker Diagram

To help clarify the Duncker diagram, let's consider the following situation. Suppose you are unhappy in your current job at your company (this is the present state—the problem to be solved) and you want to leave your company for a new job (the desired state). The left side of the diagram is the pathway for finding a new job with a different company and therefore achieving the desired state. Possible functional solutions on the left side of the diagram might include retraining for a different type of job or doing the same type of job but with another company. Specific solutions for obtaining the same type of job but with a different company might include networking to learn which companies have openings, searching the Web, and updating your résumé.

The right side of the diagram deals with alternatives that would make it okay to stay in your current job and therefore make it okay *not* to achieve the desired state of a new job. A possible functional solution for the right side of the Duncker diagram might be related to achieving greater job satisfaction. Specific solutions (that is, to have greater job satisfaction) might include greater participation in decision making, more praise, or a salary increase. This side of the diagram helps the solver get to the heart of what is really wrong in the present state and addresses that issue versus proceeding along only one of many possible solution paths to address the perceived problem.

Representing the problem as a Duncker diagram is a creative activity, and as such, there is no right way or wrong way to do it. After completing the Duncker diagram, you should try *to write a new problem statement.* This new problem statement might allow you to identify a compromise solution in which both the present state and the desired state are modified to achieve an acceptable solution. This technique can be a bit confusing at first glance. To see how it works, consider the following example of the present state/desired state technique.

We illustrate the use of the Duncker diagram with an example of a laundry detergent that could be harmful if ingested.

Bitrex

Most laundry detergents and other cleaning agents are poisonous and even fatal if consumed. Sometimes bottles of these cleaners are stored where small children can reach them and possibly drink from them. The instruction given to solve the perceived problem: *Find a way to prevent children from being able to get to the chemicals in the bottle.*

The following ideas were suggested to solve the perceived problem.

1. Print a note to adults to keep the bottle out of the reach of children. However, relying on the parents to follow these instructions is not a good idea because most adults probably will mistakenly never read them.

2. Put child-proof caps on the bottles. Again, there is a chance that the adult will be careless and will not put the cap on correctly. In addition, if the adult has trouble with the cap due to physical problems, he or she might purposely leave the cap only partially attached. The following table shows the Duncker diagram for the perceived problem statement.

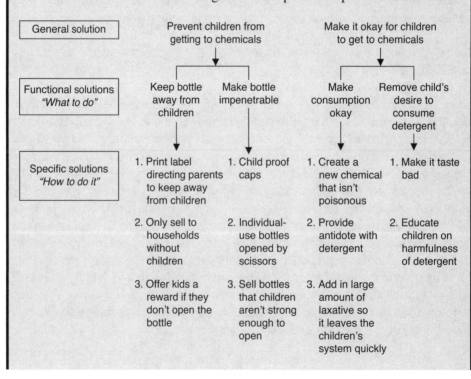

	Prevent children from getting to chemicals		Make it okay for children to get to chemicals	
General solution				
Functional solutions *"What to do"*	Keep bottle away from children	Make bottle impenetrable	Make consumption okay	Remove child's desire to consume detergent
Specific solutions *"How to do it"*	1. Print label directing parents to keep away from children	1. Child proof caps	1. Create a new chemical that isn't poisonous	1. Make it taste bad
	2. Only sell to households without children	2. Individual-use bottles opened by scissors	2. Provide antidote with detergent	2. Educate children on harmfulness of detergent
	3. Offer kids a reward if they don't open the bottle	3. Sell bottles that children aren't strong enough to open	3. Add in large amount of laxative so it leaves the children's system quickly	

The solution that one company arrived at was one much different than previously discussed. Macfarlan Smith Ltd. added a chemical that is harmless to ingest, does not affect the effectiveness of the cleaning solution, and is very bitter to the taste. Only a very small portion of this chemical placed in the solution will make it so bitter that children will not want to drink it after one taste.

New Problem Statement: Determine the exact amount of the bitter-tasting chemical to be added to the cleaning solution.

Source: *Chemical and Engineering News*, January 28, 1991, p. 41 (Matt Gdowski).

Let's consider another real-life example where the Duncker diagram uncovers the real problem and solution as opposed to the perceived problem.

Dead Rats Decaying in the Water

A new water supply system was installed in a western state. Part of the system was a large outdoor holding tank placed in the ground near a farm field. Shortly after installation, unacceptable levels of contamination were found. An investigation revealed that the contamination was caused by dead rats decaying in the holding tank. The problem statement that went out was *find a way to keep the rats out of the tank.*

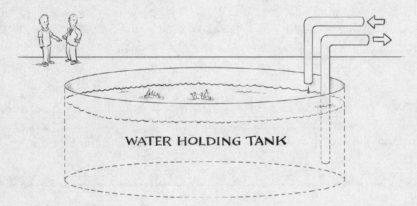

WATER HOLDING TANK

The storage tank was 200 feet in diameter and 30 feet deep, and the water level varied between one and three feet from the top of the tank. The sides of the tank were vertical and smooth and it was not known how the rats got into the tank, but they did. Because the rats are good swimmers, they would survive until exhaustion caused them to drown. Covering the tank would have been very expensive and would make collecting samples for daily testing from different areas of the tank much more difficult, so other ideas were

Continues

considered first. One proposed solution of the perceived problem was to have someone check the tank every hour to see if there are any rats swimming in the tank and then drop a bucket to pull any out. However, this solution would have been time-consuming and costly. Another solution was to always keep the tank full so rats that fell in could get out. Because of the varying water demand, keeping the tank completely full to the top would have been virtually impossible.

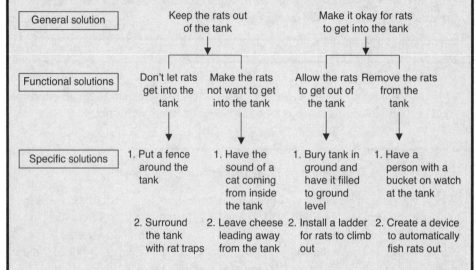

The real problem statement should have been *"Stop rats from dying and decaying in the tank."* The solution implemented was to install a ladder that allowed the rats that fell into the tank to climb out. Once the ladder was installed, dead rats were no longer found in the tank and contamination levels fell to acceptable levels.

A New Problem Statement: How and where should the ladder be placed in the storage tank and what should be the type and size of the ladder?

Source: Adapted from "Drinking Water Comes to a Boil," by Sara Terry, *The New York Times Magazine*, September 26, 1993.

TECHNIQUE 2: USING THE STATEMENT–RESTATEMENT TECHNIQUE

The **statement–restatement technique** developed by Parnes[2] is a method to evolve the problem statement to its most accurate representation of the problem. This technique is similar to the present state/desired state technique in that it requires us to rephrase the problem statement.

> *A problem well stated is a problem half solved.*
> —Charles F. Kettering

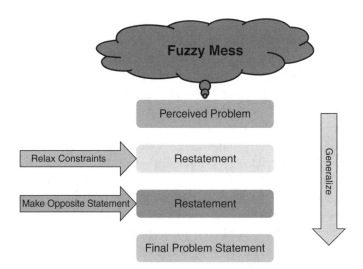

With this technique, we look at the fuzzy or unclear problem situation and write a statement regarding a problem to be addressed. The problem is then restated in different forms a number of times.

When restating the problem, it is important to inject new ideas, rather than simply changing the word order in the restated sentence. Using one or more of the following problem restatement **triggers** should prove helpful in arriving at a definitive problem statement.

Problem Statement Triggers

Trigger 1. Vary the stress pattern—try placing emphasis on different words and phrases.

Trigger 2. Choose a term that has an explicit definition and substitute the explicit definition in each place that the term appears.

Trigger 3. Make an opposite statement, change positives to negatives, and vice versa.

Trigger 4. Change "every" to "some," "always" to "sometimes," "sometimes" to "never," and vice versa.

Trigger 5. Replace "persuasive words" in the problem statement such as "obviously," "clearly," and "certainly" with the argument it is supposed to be replacing.

Trigger 6. Express words in the form of an equation or picture, and vice versa.

Each time the problem is restated, we try to generalize it further to arrive at the broadest form of the problem statement.

As an illustration of the use of these triggers, consider Trigger 3. Instead of asking, "How can my company make the largest profit?" ask, "How can my company lose the most money?" In finding the key activities or pieces of equipment that, when operated inefficiently, will result in the largest loss, we will have identified those pieces that need to be carefully monitored and controlled. This trigger helps us find the **sensitivity** of the system variables and to focus on those variables that dominate the problem.

It is often helpful to relax any constraints placed on the problem, modify the criteria, and idealize the problem when writing the restatement sentence (see trigger 4). Also, does the problem statement change when different time scales are imposed (i.e., are the long-term implications different from the short-term implications)? As we continue to restate and perhaps combine previous restatements, we should also focus on tightening up the problem statement, eliminating ambiguous words, and moving away from a fuzzy, loose, ill-defined statement.

To Market, to Market

Toasty O's was one of the first organic cereals without preservatives when it first came on the market. After several months, however, its sales dropped. The consumer survey department was able to identify that customer dissatisfaction was expressed in terms of a stale taste. The company's management then issued the following instructions to solve the perceived problem: "Streamline the production process to get the cereal on the store shelves faster, thereby ensuring a fresher product."

Unfortunately, there wasn't much slack time that could be removed from the production process to accomplish this goal. Of the steps required to get the product on the shelves (production, packaging, storage, and shipping), production was one of the fastest. Thus the company considered plans for building plants closer to the major markets, as well as plans for adding more trucks to get the cereal to market faster. The addition of either new plants or more trucks would require a major capital investment to solve the problem.

Use the statement–restatement technique to arrive at a new problem statement.

Using the Problem Statement Triggers

Original Problem Statement: The Toasty O's cereal is clearly not getting to market fast enough to maintain freshness.

Trigger 1: • **Cereal** not getting to market fast enough to maintain freshness. (Do **other products** we have get there faster?)

• Cereal not **getting** to market fast enough to maintain freshness. (Can we make the distance/time shorter?)

- Cereal not getting to **market** fast enough to maintain freshness. (Can we distribute it from a centralized location?)

- Cereal not getting to market fast enough to maintain **freshness.** (How can we keep cereal fresher, longer?)

Trigger 2:
- Breakfast food that comes in a box is not getting to the place where it is sold fast enough to keep it from getting stale. (This restatement makes us think about the box and staleness. How might we change the box to prevent staleness?)

Trigger 3:
- How can we find a way to get the cereal to market so slowly that it will never be fresh? (This restatement makes us think about how long we have to maintain freshness and what controls it.)

Trigger 4:
- Cereal is not getting to market fast enough to always maintain freshness. (This change opens up new avenues of thought. Why isn't our cereal *always* fresh?)

Trigger 5:
- The word "clearly" in the problem statement implies that if we could speed up delivery freshness would be maintained. Maybe not! Maybe the store holds the cereal too long. Maybe the cereal is stale before it reaches the store. (This trigger helps us challenge the implicit assumptions made in the problem statement.)

Trigger 6:
- Freshness is inversely proportional to the time since the cereal was baked:

$$(Freshness) = \frac{k}{(Times\ since\ cereal\ baked)}$$

This restatement makes us think about other ways to attack the freshness problem. For example, what does the proportionality constant, k, depend on?

The storage conditions, packaging, type of cereal, and other factors are logical variables to examine. How can we change the value of k?

The total time may be shortened by reducing the time at the factory, the delivery time, or the time to sell the cereal (i.e., shelf time). Once again, this trigger provides us with several alternative approaches to examine to solve the problem: Reduce the time *or* change (increase) k. Of course, the real problem was that the cereal was not staying fresh long enough—not that it wasn't reaching the stores fast enough. Keeping the cereal fresher longer was achieved by improved packaging and the use of additives to slow the rate at which Toasty O's became stale.

The new problem statement is this: "Find how to best improve packaging to keep the cereal fresher longer."

Coating Aspirin: Making an Opposite Statement

To many people, swallowing an aspirin tablet is a foul-tasting experience. A number of years ago, one company that manufactured aspirin decided to do something about it. The manager gave the following instructions to his staff to solve the perceived problem: "Find a way to put a pleasant-tasting coating on aspirin tablets." Spraying the coating on the tablets had been tried in the past, but with very little success. The resulting coating was very non-uniform, which led to an unacceptable product.

Let's apply a few triggers to this problem.

Trigger 1: Emphasize different parts of the statement.
        1. **Put** coating **on** tablet.

Trigger 3: Make an opposite statement.
        2. **Take** coating **off** tablet.

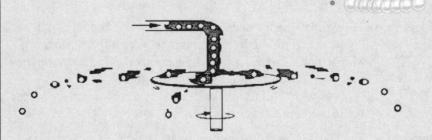

This idea led to one of the newer techniques for coating pills. The pills are immersed in a liquid in which a sweetening agent has been dissolved. The pills and liquid are then passed onto a spinning disk where the centrifugal force causes the fluid and the pills to be thrown off the spinning disk. As they are thrown, some of the liquid separates from the pill and some is held on by surface tension, leaving a nice, even coating around the pill. As the pills and liquid fall by gravity into a collecting device, the liquid evaporates, leaving the sweetening agent on the surface of the pill.

TECHNIQUE 3: KEPNER–TREGOE PROBLEM ANALYSIS

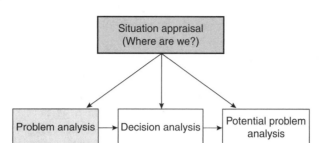

The Kepner–Tregoe[3] (K.T.) problem analysis technique is one of the four components of the K.T. approach, which includes situation appraisal, problem analysis, decision analysis, and potential problem analysis. We will discuss problem analysis in this chapter and the other three K.T. components in Chapter 8.

The K.T. problem analysis technique displays information obtained by asking critical thinking questions in a unique way called the "four dimensions of the problem." The four dimensions are what, where, when, and extent. In this technique, distinctions are made in the four dimensions as follows:

- What *is* the problem versus what *is not* the problem?
- Where does the problem occur versus where is everything okay?
- When did the problem first occur versus when was everything okay?
- What is the extent of the problem?

This kind of analysis is most useful in **troubleshooting operations** (Chapter 11) where the cause of the problem or fault is not known. Problems that lend themselves to K.T. problem analysis are ones in which an undesirable level of performance can be observed and compared with the accepted standard performance.

The basic premise of K.T. problem analysis is that there is something that distinguishes what the problem ***is*** from what it ***is not***. The cause of the problem is usually a change that has taken place and produced undesirable effects. Things were okay in the past and now they're not. Recall the example in Chapter 1 dealing with printing inks: The Bureau of Engraving and Printing changed to new printing machines and then found that ink on the new bills smeared when touched. The possible causes of the problem (deviation) were deduced by examining the differences found in the problem. In this case, the only deviation from the previous printing process was the new machines. The most probable cause of the problem is the one that best explains all the observations and facts in the problem statement. It turned out that the new machines were not putting enough pressure on the printing ink to penetrate the new paper, so it wiped off easily. The problem *was* the printing machines, *not* the printing ink.

The real challenge is to identify the distinction between the *is* and the *is not*. Particular care should be taken when filling in the "Distinction" column. Sometimes the distinction statement should be rewritten more than once to sharpen it enough to specify the distinction exactly. For example, in one problem analyzed by the K.T. method, the statement "Two of the filaments were clear (okay) and two were black (not okay)" was sharpened to "Two filaments were clear and two were covered with carbon soot." This **sharpening** of the distinction was instrumental in determining the reason for the black filament. Think in terms of dissimilarities. What distinguishes *this* fact (or category) from *that* fact (or category)? Making such a distinction requires careful analysis, insight, and practice to ferret out the differences between the *is* and the *is not*.

From the possible causes, we try to ascertain the most probable cause. The most probable cause is the one that explains each dimension in the problem specification. The final step is to verify that the most probable cause is the true cause. This step may be accomplished by making the appropriate change to see if the problem disappears.

The problem solver should also separate people's observations from their interpretations of what went wrong. A common mistake is to assume that the most obvious conclusion or the most common is always the correct one. (This point is a good place to start, but not necessarily to stop.) A famous medical school proverb that relates to the diagnosis of disease is this: "When you hear hoof beats, don't think zebras." In other words, look for simple explanations first. Finally, the problem solver should continually reexamine the assumptions and discard them when necessary.

> *When you hear hoof beats, don't think zebras!*

> *Is, or is not? That is the question, Watson!!!*

Dimec/Shutterstock

The Four K.T. Dimensions of a Problem

		Is	**Is Not**	**Distinction**	**Cause**
What	Identify	What is the problem?	What is not the problem?	What is the distinction between the **is** and the **is not?**	What is a possible cause?
Where	Locate	Where is the problem found?	Where is the problem not found?	What is distinctive about the difference in locations?	What is the possible cause?
When	Timing	When does the problem occur?	When does the problem not occur?	What is distinctive about the difference in the timing?	What is a possible cause?
		When was the problem first observed?	When was the problem last observed?	What is the distinction between these observations?	What is a possible cause?

		Is	Is Not	Distinction	Cause
Extent	Magnitude	How far does the problem extend?	How localized is the problem?	What is the distinction?	What is a possible cause?
		How many units are affected?	How many units are not affected?	What is the distinction?	What is a possible cause?
		How much of any one unit is affected?	How much of any one unit is not affected?	What is the distinction?	What is a possible cause?

Copyright © Kepner–Tregoe, Inc., 1994. Reprinted with permission.

Fear of Flying

A new model of airplane caused an unexpected issue when delivered to a major airline. Immediately after the planes were put into operation, many of the flight attendants developed a red rash on their arms, hands, and faces. The rash did not appear on any other part of the body, and it occurred only on flights that went over water. Fortunately, the rash usually disappeared in 24 hours and caused no additional problems beyond that time. When the attendants flew old planes over the same routes, no ill effects occurred. The rash did not occur on all the attendants of a particular flight, but the same number of attendants contracted the rash on each flight. In addition, those flight attendants who contracted the rash felt no other ill effects.

Continues

The flight attendants' union threatened action because the attendants were upset and worried, and the attendants believed that some malicious force was behind the mysterious rash. Many doctors were called in, but all were in a quandary. Industrial hygienists could not measure anything extraordinary in the cabins

Source: *Chemtech*, 13, 11, p. 655, 1983. Image: idiz/Shutterstock.

Let's apply K.T. problem analysis to this situation to see if we can learn the cause of the problem.

What

The problem appears to be only a rash and not any other illness such as headache or nausea. A rash can be caused by external contact with an allergen such as poison ivy. Consequently, we can complete the first row of the K.T. table as follows:

	Is	Is Not	Distinction
What	Rash	Other illness	External contact

When

The problem occurs only when the new planes are used, but not when the old planes are used. The new planes contain new materials such as lighter composite materials and new fabric, among other things.

	Is	Is Not	Distinction
When	New planes used	Old planes used	New materials

Where

The illness occurs only in flights over water; attendants on flights over land do not develop the rash. Different crew procedures are followed on flights over water than are followed on flights over land.

	Is	Is Not	Distinction
Where	Flights over water	Flights over land	Different crew procedures

Extent

There are two categories of "extent" in this case: (1) how many units are affected and (2) how much of any one unit is affected. In terms of how many units (flight attendants) are affected, only some of the attendants are affected;

not all are affected. The different attendants have different duties. How much of any one unit (attendant) is affected is defined as the fact that the rash appears only on the face, hands, and arms. The culprit must be something contacting the face, hands, and arms.

	Is	Is Not	Distinction
Extent	Only some of the attendants	All of the attendants	Different crew duties
Extent	Face, hands, and arms	Other body parts	Something contacting exposed face, hands, and arms

Let's summarize what we have learned so far:

	Is	Is Not	Distinction
What	Rash	Other illness	External contact
When	New planes used	Old planes used	New materials
Where	Flights over water	Flights over land	Different crew procedures
Extent	Only some of the attendants	All of the attendants	Different crew duties
	Face, hands, and arms	Other body parts	Something contacting exposed face, hands, and arms

When we look at all of the distinctions, we see that (1) something contacting the arms and face could be causing the rash; (2) the rash occurs only on flights over water, and the use of life vests is demonstrated on flights over water; and (3) the life vests on the new planes are made of new materials or of a different brand of materials and usually two flight attendants demonstrate the use of

Continues

these life vests. The new life preservers contained some material that was the rash-causing agent!

A Pre-flight Demo of a Life Vest

Photo: Serg Zastavkin/Fotolia.

Spontaneous Car Starting

Mikhail Bakunovich/
Shutterstock

Molly traded in her old car for a one-year-old, high-end car that had a special keyless entry system. Everything about the car worked perfectly, with one exception: When Molly would park the car, sometimes she would come out to find its motor running. In each instance, she had turned the motor off and taken the key out before leaving the car. This problem most often occurred at the Starbucks on Washtenaw Avenue, a major thoroughfare, but seldom at the Starbucks on Liberty Street, which was a more secluded street. It also occurred a couple of times at Dunkin' Donuts on Stadium Street. Even though the problem mainly occurred on Washtenaw, Molly did not want to stop going there because she enjoyed chatting with the police officers that frequented that location. The Liberty Street Starbucks seemed to have only an occasional police officer stop by. The problem arose during the midmorning hours but not over lunch or dinnertime.

	Is	Is Not	Distinction
What	Spontaneous starting	Regular starting	Signal to start car

	Is	Is Not	Distinction
Where	Starbucks on Washtenaw and Dunkin' Donuts on Stadium	Starbucks on Liberty	Police in Washtenaw Starbucks and Dunkin' Donuts

	Is	**Is Not**	**Distinction**
When	Midmorning	Lunch	Police taking morning break
	New car	Old car used	New cars have remote entry and ignition system

	Is	**Is Not**	**Distinction**
Extent	Only in locations that the police frequent	Locations where there are no police	Police doing something like receiving/sending calls

The newer-model keyless entry system had a remote starting capability when a signal was sent from the key chain. The frequency from the police radios matched the frequency from the key chain that started the engine.

Source: From an idea on the National Public Radio show *Car Talk*.

After the problem is defined, there is more work to be done. Nevertheless, having a well-defined problem puts us well on the way toward finding a solution.

WHICH TECHNIQUE TO CHOOSE?

We do not expect you to apply every problem definition technique to every situation. In fact, when 400 problem solvers were surveyed as to which two techniques presented in this chapter were the most useful to them, their choices were virtually equally divided among those presented in this chapter. In other words, different techniques work better for different individuals and different situations, and your selection of a problem definition technique is a matter of personal choice. The main point is to be organized as well as creative in your approach to problem definition.

DETERMINE WHETHER THE PROBLEM SHOULD BE SOLVED

Having defined the real problem, we now need to develop criteria by which to judge the solution to the real problem. One of the first questions experienced engineers ask is this: Should the problem be solved? The figure following this discussion shows how to proceed when answering this question. As you can see from the figure, the first step is to determine whether a solution to an identical problem or a similar problem is available. A literature search may determine whether a solution exists.

Establish criteria to judge the solution.

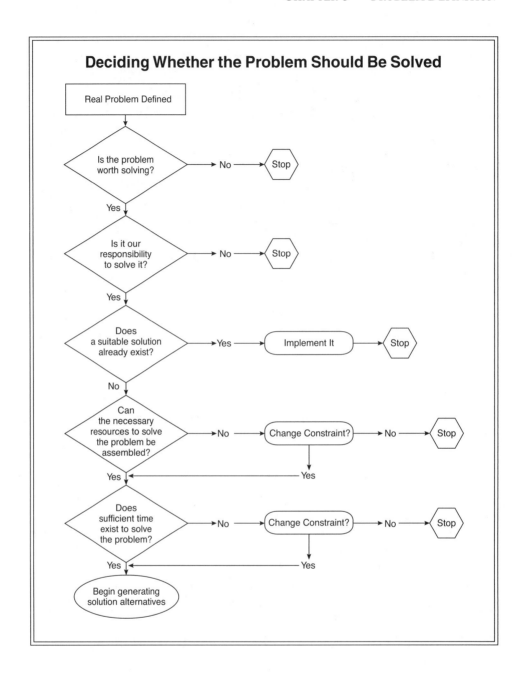

Deciding Whether the Problem Should Be Solved

How do experienced engineers go about deciding whether the problem is *worth* solving? Perhaps the problem is just mildly irritating and consequently may be ignored altogether. For instance, suppose the garage door at your plant's warehouse facility is too narrow to allow easy access by some of the delivery vehicles. They can pass through, but the clearance is very tight. This is an annoying problem, but if the fix is quite costly, you could probably "live with it."

Here are some questions you should ask early in the process:

- What are the resources available to solve the problem?
- How many people can you allocate to the problem, and for how long?
- How soon do you need a solution? Today? Tomorrow? Next year?

These are key questions to keep in mind as you take your first steps along the way to finding a problem solution. The quality of your solution is often—but not always—related to the time and money you have to *generate it and carry it through*. In some instances, it may be necessary to extend deadlines to obtain a quality solution.

It may not be possible to completely address the cost issue until we are further along in the solution process. The cost will depend on whether the solution will be a permanent one or a temporary or patchwork solution. Sometimes *two* solutions are required: one to treat short-term symptoms so as to keep the process operating and one to solve the real problem for the long term. Be aware of these two mindsets in the problem-solving process. In some cases the "No" answers in the above figure can be changed to "Yes" by *selling* the project to management so as to get additional resources or extend the deadline. This change can be achieved by showing that the problem is an important one and is relevant to the operation of the company.

Let's apply this algorithm to the example of the dead fish discussed in Chapter 3. "Is the problem worth solving?" The answer is yes, because fungi will continue to affect more fish if we don't do something. Therefore we continue to the next step: "Is it the company's responsibility to solve the problem?" The answer to this question is **no** because the company did not cause the problem and it is the responsibility of the state's Department of Natural Resources (DNR). They should not have the obligation to cure the fish or stop the fungi from spreading just because they were accused of causing the problem. Consequently we should *stop* our problem-solving process.

BRAINSTORM POTENTIAL CAUSES AND SOLUTION ALTERNATIVES

This brings us to the end of the first phase of the creative problem-solving process and to the first step of the second phase of the process: generating solutions to problems. Techniques to generate solutions will be discussed in Chapters 6 and 7.

SUMMARY

In this chapter, we discussed the necessity for defining the real problem. Three problem definition techniques were presented to help you zero in on the true problem definition.

- **Duncker Diagram**
  - Analyze two pathways: (1) a pathway to your desired state and (2) a pathway that makes it okay not to achieve the desired state.

Continues

- **Statement–Restatement**
  - Use the six triggers to restate the problem in a number of different ways.

- **K.T. Problem Analysis**
  - What is? What is not?
  - Where is? Where is not?
  - When is? When is not?
  - Extent is? Extent is not?

WEB-SITE MATERIAL (WWW.UMICH.EDU/~SCPS)

- **Learning Resources**
  Summary Notes
  Thoughts on Problem Solving (access through the home page)

- **Interactive Computer Modules**
  Thoughts on Problem Solving (access through the home page)

Problem Definition 1

Problem Definition 2

- **Professional Reference Shelf**
  1. Exploring the Problem
     - The Case of the Dead Fish
     - A Real Tooth Ache
     - De-bottlenecking a Process (Heat Exchanger)

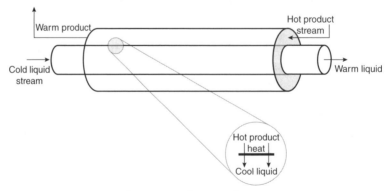

Refrigerator (heat exchanger)

2. Present State/Desired State
   – Missing the Mark
3. Duncker Diagrams
   – Kindergarten Cop
   – To Market, to Market
4. Statement–Restatement Techniques
   – Exotic Materials
   – Arid Land
   – Leaking Flow Meter
   – Wake Up and Smell the Coffee
5. Closed-Ended Algorithm
   – Chemical Reaction Engineering and more

REFERENCES

1. Higgins, J. S., et al., "Identifying and Solving Problems in Engineering Design," *Studies in Higher Education*, 14, 2, p. 169, 1989.
2. Parnes, S. J., *Creative Behavior Workbook*, Scribner, New York, 1967.
3. Kepner, C. H., and B. B. Tregoe, *The New Rational Manager*, Princeton Research Press, Princeton, NJ, 1981.

EXERCISES

5.1. Load and run "Interactive Computer Module Problem Definition 1" from the SCPS Web site. Record your performance number: _____.

5.2. Load and run "Interactive Computer Module Problem Definition 2" from the SCPS Web site. Record your performance number: _____.

5.3. Make a list of the most important things you learned from this chapter. Identify at least three techniques that you believe will change the ways you think about defining and solving problems. Which problem definition techniques do you find most useful? Prepare a matrix table listing all of the problem definition techniques discussed in this chapter. Identify those attributes that some of the techniques have in common and those attributes that are unique to a given technique.

	Attribute 1	Attribute 2	Attribute 3
Technique A	X		X
Technique B		X	X
Technique C	X	X	

5.4. Write a sentence describing a problem you have. Apply the triggers in the statement–restatement technique to your problem.

Perceived Problem Statement: _____

Restatement 1: _____

Restatement 2: _____

Final Problem Statement: _____

Next apply the Duncker diagram to the same problem.

5.5. Carry out a present state/desired state analysis and prepare a Duncker diagram for the following problem: "I want a summer internship but no one is hiring."

5.6. You have had a very hectic morning, so you leave work a little early to relax a bit before you meet your supervisor, who is flying in to a nearby airport. You have not seen your supervisor from the home office for about a year now. He has written to you saying that he wants to meet with you personally to discuss the last project. Through no fault of yours, everything went wrong: The oil embargo delayed shipment of all the key parts; your project manager had a skiing accident; and your secretary enclosed the key files in a parcel that was sent by mistake to Japan via sea mail. Your supervisor thinks that you have been so careless on this project that you would lock yourself out of your own car.

As you are driving through the pleasant countryside on this chilly late fall afternoon, you realize that you will be an hour early. You spot a rather secluded roadside park about 200 meters away. A quiet stream bubbles through the park, containing trees in all their autumn colors—such an ideal place to just get out and relax. You pull off into the park, absentmindedly get out and lock the car, and stroll by the stream. When you return, you find that you have locked the keys inside the car. The road to the airport is not the usual route; cars pass through about every 10 to 15 minutes. The airport is nine kilometers away; the nearest house (with a telephone) is one kilometer away. The plane is due to arrive in 20 minutes. Your car, which is not a convertible, is such that you cannot get under the hood or into the trunk from the outside. All the windows are up and secured.

Apply the Duncker diagram and one other problem-solving technique to help you decide what to do.

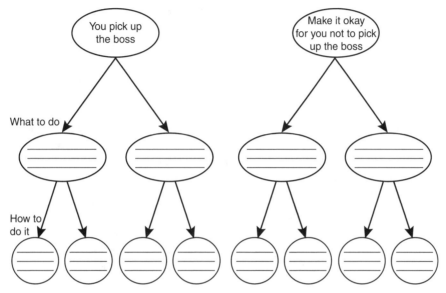

Problem courtesy of D. R. Woods, McMaster University.

5.7. Sodium azide is used as the propellant in an air bag system. This chemical is mixed with an oxidizing agent and pressed into pellets, which are then hermetically sealed into a steel or aluminum can. Upon impact, ignition of the pelletized sodium azide generates nitrogen gas that inflates the air bag. Unfortunately, if the sodium azide contacts acids or heavy metals (e.g., lead, copper, mercury, or their alloys), it forms toxic and sensitive explosives. Consequently, at the end of an automobile's life, a serious problem surfaces when an automobile with an undetonated air bag is sent to the junkyard for compacting and shredding, because it might contact heavy metals during this process. The potential for an explosion during processing represents a serious danger for the people who operate the scrap recycling plant. Apply two or three problem definition techniques to this situation. Source: *Chemtech*, 23, p. 54, 1993.

5.8. Pillsbury, a leader in the manufacture of high-quality baking products, had its origins in the manufacture of flour for the baking industry. At the time Charles Pillsbury purchased his first mill in Minneapolis, the wheat from Minnesota was considered to be substandard compared to the wheat used in the mills in St. Louis, which was then the hub of the milling industry. Part of the problem was that winter wheat, which is commonly used in high-grade flour, could not be grown in Minnesota because of the long and cold winters there. Consequently, the Minnesota mills were forced to use spring wheat, which had a harder shell. At the time, the most commonly used milling machines used a "low grinding" process to separate the wheat from the chaff. The low grinding process refers to the use of stone wheels. A stone wheel rests directly on the bottom wheel, with the wheat to be ground placed between them. With harder wheats, a large amount of heat was generated during the grinding process, discoloring and degrading the product's quality. Thus the flour produced from the Minnesota mills was discolored and inferior, and it had less nutritional value and a shorter shelf life than flour produced in the St. Louis mills. The directions given could have been "Order more river barges to ship winter wheat up the Mississippi River from St. Louis to Minneapolis." Apply two or more problem definition techniques to this situation.

Source: Adapted from "When in Rome" by Jane Ammeson, *Northwestern Airlines World Traveler*, 25, 3, p. 20, 1993.

5.9. Detroit, Michigan, and Windsor, Ontario, are separated by the Detroit River and a distance of approximately one kilometer. Off and on, over the last two years, the residents of Windsor have complained of an annoying noise that sounds like a loud diesel truck idling, metallic grating, or a loud boombox coming from the American side. They have said this hum, that is, noise pollution, is causing illness, whipping dogs into frenzies and disrupting nocturnal baby feedings. The hum is not heard all the time in Canada and has never been heard on the American side of the river. After three months of seismic study, the Canadian government said it was likely the noise was coming from Zug Island in the middle of the Detroit River. A U.S. official said that on over 20 recent trips to the island he never heard the noise. However, hundreds of sleep-deprived locals have complained so frequently that Canada's parliamentary secretary wants the United States to pay for the investigation that will help quiet down the border ruckus. U.S. officials have refused, saying that there have not been any complaints in the Detroit area. Suggest a path forward.

Dan Janisse/*Windsor Star*

Sources: Alistair MacDonald and Paul Vieira, "Canadians Make a Racket over Mysterious 'Windsor Hum': Unexplained Noise Spurs Diplomatic Fracas at Detroit Border; Americans Can't Hear It," *Wall Street Journal*, April 30, 2012; Elizabeth Chuck, "Mysterious Booms in Wisconsin Town," NBC News, March 21, 2012.

5.10. An airline at the Houston Airport tried to please its passengers by always docking its planes at gates within a one- to two-minute walk to the airport entrance and baggage claim and by having all the bags at the baggage claim area within eight to ten minutes. Unfortunately, the airline received many complaints about the time it took to get the bags to the baggage claim area. The airline researched the situation and found that there was virtually no way it could unload the bags to the transport trucks, drive to the unloading zone, and unload the bags any faster. The airline didn't change the baggage unloading procedure, but it did change another component of the arrival process and the complaints disappeared. The airline did not use mirrors to solve the problem (as was the case for the slow elevators).

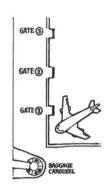

A. What was the real problem?

B. Suggest a number of things that you think the airline might have done to eliminate the complaints. Apply two or more problem-solving techniques.

Source: *Washington Post*, p. A3, December 14, 1992.

5.11. In 1991, 64% of all commercial radio stations in the United States lost money. For a radio station to remain solvent, it must have a significant revenue stream from advertisers. Advertisers, in turn, target the markets they consider desirable (i.e., based on listeners' income, spending, and interests), and for the past several years this target has been the age group from 25 to 54. Along with the revenue loss, the number of radio stations playing the Top 40 songs (i.e., the 40 most popular songs of that week) has decreased by a factor of 2 in the past three years, as did the audience for the Top 40 songs. Many stations tried playing a blend of current hits and hits from 10 to 20 years ago; however, this blend irritated the younger listeners and did not seem to solve the economic problem. Apply two or more problem definition techniques to this situation.

NKLRDVC/Shutterstock

Source: Adapted from *The International Herald Tribune*, p. 7, March 24, 1993.

5.12. FireKing is a small manufacturer of rich-looking fireproof filing cabinets that wants to increase its current market share of 3%. While the company's designs are elegant, the cabinets are also the heaviest ones on the market—and in customers' minds, greater weight means higher quality. However, greater weight also means higher shipping and transportation costs, which makes the cabinets very expensive. FireKing's management asks the following question: "How can we make our product lighter so as to have a competitive price?" However, some executives believed a lighter-weight product might hurt the image of quality. Apply one or more problem definition techniques to this situation.

Problem courtesy of David Turczyn.

5.13. Bug-B-Gone Company has developed a new method for killing roaches that is more effective than any of the other leading products. In fact, no spraying is necessary because the active ingredient is held in a container that is placed on the floor or in corners; the roach problem disappears when the container is activated. This method has the advantage that the product does all the work—the user does not need to search out and spray the live roaches. The product was test-marketed to homemakers in some southern states. Everyone who saw the effectiveness test results agreed that the new product was superior in killing roaches. Despite a massive advertising campaign, however, the standard roach sprays are still far outselling the new product. Apply one or more problem definition techniques to this situation.

Problem courtesy of David Turczyn.

5.14. A major American soap company carried out a massive advertising campaign in an attempt to expand its sales in Poland. The television commercials featured a beautiful woman using the company's soap during her morning shower. Thousands of sample cakes were distributed door to door throughout the country. Despite these massive promotional efforts, the campaign was entirely unsuccessful. Polish television had been used primarily for Communist Party politics, and commercials were relatively rare. What was aired on television was usually party-line politics. Apply one or more problem definition techniques to this situation.

Problem courtesy of Christina Nusbaum.

5.15. Employees of a certain store are allowed to take merchandise out of the department store on approval. The original procedure required each employee to write an approval slip identifying the merchandise taken. Some employees are abusing the system by taking the clothing and destroying the slip, thereby leaving no record of the removed merchandise. Apply one or more problem definition techniques to this situation.

Problem courtesy of Maggie Michael.

5.16. After Crest toothpaste had been on the market for some time, Procter & Gamble, its manufacturer, decided to offer a mint-flavored version in addition to the original, wintergreen-flavored product. In the course of developing the new mint-flavored product, a test batch of mint-flavored product was produced by the same pilot unit used to produce wintergreen-flavored product. The pilot equipment used a tank and impeller device to mix the mint-flavor essence with the rest of the ingredients to form the finished product (which is a very viscous solution). Some of the pilot plant product was packed into the familiar collapsible tubes for further testing. Tubes used in testing the mint flavor were identical to those used for the wintergreen-flavored product. In the packing operation, toothpaste was pumped through lines into the unsealed ends of brand-new tubes. After filling, the open tube ends were heatsealed. The packing operation is illustrated in the figure.

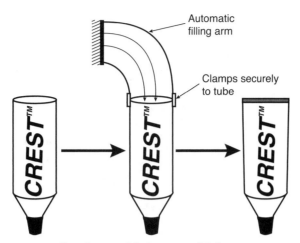

Automatic
filling arm

Clamps securely
to tube

Continuous Movement of Tubes

To assess storage stability, some of the filled tubes were randomly separated into several groups and each group was stored in a constant-temperature room. Storage temperatures varied from 40°F to 120°F.

Early sampling of the mint-flavored product showed nothing unusual. However, several months into the test, a technician preparing to test the product from one of the stored tubes noted that the first 0.25 inch of paste squeezed onto a toothbrush was off-color. The rest of the product in the tube met the color specification. Nothing like this had ever been seen with the original formula. Further testing showed that a person had to squeeze more product out of those tubes that had been stored at higher temperatures and/or stored for longer times before a product that met the color specifications would exit the mouth of the toothpaste tube. Tubes stored for a period of time at 40°F contained no off-color product, whereas tubes stored for the same length of time at higher temperatures produced off-color paste.

The only exception to these results was a single tube, which had been stored at a temperature above 40°F. A leakage of off-color product was found around the base of the cap on this tube, but the product inside the tube met the color specifications.

While other tests showed the off-color product to be safe and effective in cleaning teeth, consumers certainly would not accept a color change in a product expected to have the same color from the first squeeze to the last. Moreover, such a change could have been an early warning of more serious problems to come. This phenomenon had to be understood and eliminated before the new flavor could be marketed. Accordingly, various possible remedies were tested: caps and tubes made of different materials, different mixing methods, and so forth. None of these changes had any effect on the off-color problem. All raw materials, including the new mint-flavor essence, were checked and found to meet specifications. A subsequent batch of the wintergreen product was made and tested for storage stability and, as usual, no off-color problems occurred.

Carry out a K.T. problem analysis to learn the cause of the off-color toothpaste.

5.17. Chocolate butter paste is the primary ingredient used by a number of major bakeries for a wide variety of pastries. The paste is a very viscous liquid that is manufactured by Cocomaker Industries in a major populous city in the Midwest. Cocomaker supplies customers as close as Dolton and as far away as Chicago (which is a long drive from its plant). The paste flows from the production line into five-gallon drums, which are placed immediately into refrigerated trucks for shipment to the respective customers. Until February, all the trucks had been the same size and the drums were stacked in rows three drums wide, four drums high, and eight drums deep. Now two rather small customers, Bell Bakery and Clissold Bakery, each requiring 20 drums per day, have been added in the Chicago area. Supplying these new customers, along with an increased order by the Chicago customer Hoyne Industrial Bakers, necessitated the purchase of a larger truck. The new truck could fit five drums across, four drums high, and eight drums deep. The truck would stop at Bell, then Hoyne, then Clissold in Chicago proper. With the increased market in the Chicago area, Cocomaker's plant is running at close to maximum capacity. Because the ingredients of the paste are mixed by static mixers, the pumps are currently operating at their maximum capacity and the plant is operating 20 hours per day.

In November, Cocomaker successfully lured two nearby customers, Damon Bakery and Oakley Bakery, away from one of its competitors. By increasing plant operation to 24 hours per day, all orders could be filled. As the Christmas season approached, however, the usual seasonal demand for chocolate butter paste posed a problem for Cocomaker: It needed to meet demands not encountered in previous years. The company decided that if the processing temperature was increased by 20 degrees, the paste would be sufficiently less viscous, and production demands could be met with the current pump limitations. Unfortunately, the increased capacity began to generate problems as Christmas approached. The pumps began failing on a regular basis; a strike at the supplier of the shipping containers caused Cocomaker to buy containers from a new supplier, which claimed to carry only sturdier containers at a 10% increase in price; the safety officer had an emergency appendectomy; and—most troubling—Hoyne called about an unacceptable bacteria count in shipments for the last five days. As a result of the bacteria, people who bought Hoyne's products have been getting ill.

An immediate check of the bacteria levels shows that they are at the same acceptable levels they have always been when leaving Cocomaker's plant. You call Hoyne and tell them that the plant levels show that the paste is within bacteria specifications. Two days later, you receive a call from Hoyne, saying that the bakery had hired an independent firm, which reported that the bacteria levels for the chocolate butter paste are well above an acceptable level. You call the Damon, Bell, Clissold, and Oakley bakeries and ask them to check their bacteria counts; they report back that everything is within the specifications most often reported. A spot check of other customers shows no problems. Now you receive a call from Hoyne, saying the bakery is initiating legal and governmental actions to close your plant down.

Carry out a K.T. problem analysis to learn the cause of the problem.

5.18. "Blue and Green Honey Makes French Beekeepers See Red"

Bee graphic: Pagina/Shutterstock; honeycomp photo: Africa Studio/Shutterstock

France is one of the largest producers of honey within the European Union, producing some 18,330 tons annually, according to a recent audit conducted for the national farm agency FranceAgriMer.

Ribeauville, situated in the middle of Alsace along the scenic wine route southwest of Strasbourg, is best known for its vineyards. The economic situation in the region is excellent, nesting great industries and factories such as Aventis, Brasseries Kronenbourg, Mars Chocolate, and Agrivalor side-by-side with the vineyards. Aventis is a pharmaceutical company manufacturing Ambien and anticancer and other drugs. Kronenbourg is a brewery whose beer is distributed worldwide. Mars Chocolate produces candy such as Mars Bars, Milky Way, Snickers, Skittles, Starburst, M&Ms, Dove Bars, 3 Musketeers, and Twix. Agrivalor is a biogas company that takes waste from these industries and turns it into a gaseous fuel. In addition, living among the winemakers are about 2,400 beekeepers in Alsace who tend some 35,000 colonies and produce about 1,000 tons of honey per year, according to the region's chamber of agriculture.

The continuous success of the region's economy is partially due to the intermittent participation of the local government in creating incentives to develop integration among the local companies. Recently the local authorities have launched a tax program that gives a tax credit for companies that locally integrate their supply chains, including raw materials and waste disposal. As a result Kronenbourg and Mars have been sending their waste to a biogas plant. Each of Mars's products is sent separately in different containers to the plant while Kronenbourg combines all its waste together and ships it in covered barrels.

Recently, bees at a cluster of apiaries, also known as bee yards, in northeastern France have been mysteriously producing honey in shades of blue and green, alarming their keepers. The unsellable honey is a new headache for around a dozen affected beekeepers already dealing with high bee mortality rates and dwindling honey supplies following a harsh winter, said Alain Frieh, president of the beekeepers' union.

Since August, beekeepers around the town of Ribeauville have seen bees returning to their hives carrying unidentified colorful substances that have turned their honey unnatural shades of brown, yellow, blue, and green. Mystified, the beekeepers embarked on an investigation and discovered that the biogas plant was only four kilometers (2.5 miles) away from the apiaries. Because of the unsellable colored honey, the beekeepers decided to call a meeting with the other members of their union. Since the use of pesticides was banned, it was not a reason to be worried. Some of the members of the union brought up the possibility of contamination from the industries in the region. Thus, they perceived that the colored honey was only in the proximities of the biogas plant exclusively. However, the problem only started last August and the biogas plant has been installed there for much longer. The biogas company had to shut down for a couple of weeks in August due to malfunctioning equipment and as a result had to store the open waste containers containing residue from the different products from the Mars factory in an uncovered holding area. The waste from Kronenbourg was stored beside the Mars waste; however, it remained in covered barrels.

Bee numbers have been rapidly declining around the world in the last few years and the French government has banned a widely used pesticide, Cruiser OSR, that one study has linked to high mortality rates.

Carry out a K.T. problem analysis to learn the cause for the colored honey.

Source: Blue and Green Honey Makes French Beekeepers See Red by Patrick Genthon. *Reuters*, Oct. 5, 2012. © Thomson Reuters 2012. All rights reserved.

5.19. Sparkling mineral water is the primary product of Bubbles, Inc. This firm, which is based in France, serves three major markets—Europe, North America, and Australia. It collects water from a natural spring; the water is then filtered through a parallel array of three filter units, each containing two charcoal filters. The filtration process removes trace amounts of naturally occurring contaminants. The filtered water is stored in separate tank farms, one for each market, until it is transported by tanker truck to one of the three bottling plants that serve the company's markets. When the water arrives at the bottling plant, it is temporarily placed in 3500 m^3 storage tanks until it can be carbonated to provide the effervescence that is the trademark of the producer. Some of the water is also flavored with lemon, cherry, and raspberry additives.

Next, the sparkling water is packaged in a variety of bottle sizes and materials, ranging from 10-ounce glass bottles to 1-liter plastic bottles. The European market receives its shipments directly by truck, usually within three days. Products bound for North America or Australia are shipped first by truck to the waterfront and then by freighters to their overseas destinations.

Business has been good for the last several months, with the North American and European markets demanding as much sparkling water as can be produced. This situation has required that Bubbles contract with additional plastic bottle suppliers to keep up with the increased demand. It has also forced regularly scheduled maintenance for the Australian and North American tank farms to be delayed and rescheduled because of the high demand for the product. There is also, of course, a larger demand placed on the spring that supplies the mineral water for the process.

Unfortunately, the news is not all good for Bubbles. The bottling plant for the Australian market is currently several weeks behind schedule owing to a shipment that was lost at sea. This catastrophe has required that water from the company's reserve springs, which are located many miles from the bottling plant, be used to augment the water supplied by the regular spring so that the bottling plant can operate at an even higher level of production. The availability of water from the reserve springs is hindered by their remote locations, but the water from these springs does not require filtration. In addition, contract negotiations are going badly and it appears there will be a strike at all of the bottling plants. Recent weather forecasts indicate that relief from the ongoing drought, which has already lasted three months, is not likely. Worst of all, customers in the North American and Australian markets are complaining that all shipments of the sparkling water in the last six weeks have contained benzene in unacceptably high concentrations. You know that benzene is often used as an industrial solvent but is also found naturally.

A quick survey of the bottling plant managers shows that the North American–bound products that are currently packaged and awaiting shipment have benzene concentrations in excess of acceptable concentrations. However, the managers of the bottling plants that service the Australian and European markets report that no significant level of benzene was detected in the bottles that are currently stored. Authorities in the North American and Australian markets have already begun recalling the product, with authorities in the European market pressuring

LIVERPOOL JOHN MOORES UNIVERSITY
LEARNING SERVICES

Bubbles for a quick solution and threatening to recall products as a precautionary measure.

Carry out a K.T. problem analysis to learn the cause of the problem.

Source: Adapted from *Chemtech*, "When the Bubble Burst," p. 74, February 1992.

5.20. Currently, many drilling platforms in the Gulf of Mexico collect oil from a number of oil wells, which they then pump through a single pipeline from the platform to shore. Most of these wells have always been quite productive. Consequently, the oil flows through the pipeline lying on the ocean floor at a reasonable rate.

When the oil comes out of the wells, it is at a temperature of approximately 145°F; by the time the oil reaches shore, the temperature in most pipelines is approximately 90°F. The temperature of the water on the ocean floor for the majority of the platforms within two miles of shore is approximately 45°F. However, the water depth increases as you move away from shore and the temperature of the water on the ocean floor decreases to 39°F.

Recently, two new platforms (A and B) were erected in the Gulf Coast farther out from shore than the other drilling platforms. About a year and a half after both came on stream, a disaster occurred on Platform A, such that no oil could be pumped to shore through the pipeline from this platform. Platform B continued to operate without any problems. When the crude composition at the well head was analyzed, it was found to have exactly the same chemical composition (e.g., asphaltenes, waxes, gas) as that found in the well heads on all other platforms. The only difference between Platform A and Platform B was that the production rate of Platform A was much lower than that of Platform B. Nevertheless, the production rate from Platform A was still greater than many of the platforms near the shoreline.

Carry out a K.T. problem analysis to learn the reason for the plugging of pipeline A.

5.21. Research or create a statement of a perceived problem and then apply one or more of the following problem definition techniques to that problem statement:

A. Find out where the problem came from

B. Present state/desired state and Duncker diagram

C. Statement–restatement technique

D. Problem analysis (*Hint*: See the example and homework problems in Chapter 12.)

5.22. Dow Chemical Company, a world leader in supplying special chemical products and solutions to all sectors, launched a new product named Dursban, a pesticide specifically for use in Latin American climates. Dow exports significant amounts of pesticide to Latin America where the industrial farms make up 75% of sales and the other 25% are to small individual farmers. Not long after the launch of Dursban, Dow received complaints from some of the large industrial farms that a strong odor was emitted from their fields. They blamed the product Dursban for causing these strong odors because the odor began when they switched to the new pesticide. The management team of engineers at the chemical company decided to install a new multimillion-dollar crystallization separation process at the manufacturing plant in America, in September, to improve the pesticide and to get rid of the odor. The new odor-free Dursban was released later the same year in November. The engineers tested the product thoroughly to make sure the new version was just as effective as the previous version as the initial results showed, and it was. The industrial farms in Latin America praised the new product as their crops were doing just as well with the new product as the old and there was no odor. The company was pleased with its new product and even applied the technology to its North American pesticide to reduce its odors to the applause of Dow's large American customers, which are nearly all large industrial operations. To Dow's surprise, sales of the new product decreased by 25% in Latin America by the beginning of the next year.

Carry out a K.T. problem analysis to find the real problem of why sales of the new product decreased in Latin America.

Source: Steve Pankratz, University of Michigan, BSE Chemical Engineering, Class of 2010.

FURTHER READING

Copulsky, William. "Vision Innovation." *Chemtech*, 19, p. 279, May 1989. Interesting anecdotes on problem definition and vision related to a number of popular products.

De Bono, Edward. *Serious Creativity*. Harper Business (a division of Harper Collins), New York, 1993. Summary of 20 years of creativity researched by de Bono. Many useful additional problem definition techniques are presented.

Rubenfeld, M., and B. Scheffer. "Critical Thinking: What Is It and How Do We Teach It?" In: J. Dochterman and H. Y. Grace (eds.), *Current Issues in Nursing*. Mosby, St. Louis, MO, 2001, pp. 125–132.

Scheffer, B. K., and M. G. Rubenfeld. "A Consensus Statement on Critical Thinking in Nursing." *Journal of Nursing Education,* 39, pp. 352–359, 2000.

The following books are all published by the Foundation for Critical Thinking (P.O. Box 220, Dillon Beach, CA 94929, www.critical thinking.org):

Linda Elder and Richard Paul. *The Miniature Guide to Analytic Thinking* (2003).

Linda Elder and Richard Paul. *The Miniature Guide to Asking Essential Questions* (2002).

Richard Paul and Linda Elder. *The Miniature Guide to Critical Thinking Concepts and Tools* (2009).

Richard Paul and Linda Elder. *The Thinker's Guide to Critical and Creative Thinking* (2004).

Richard Paul and Linda Elder. *The Thinker's Guide to the Art of Socratic Questioning* (2006).

Richard Paul, Robert Niewoehner, and Linda Elder. *The Miniature Guide to Engineering Reasoning* (2006).

6 BREAKING DOWN THE BARRIERS TO GENERATING IDEAS

For every failure, there's an alternative course of action. You just have to find it. When you come to a roadblock, take a detour.

—Mary Kay Ash
(Founder of Mary Kay Cosmetics)

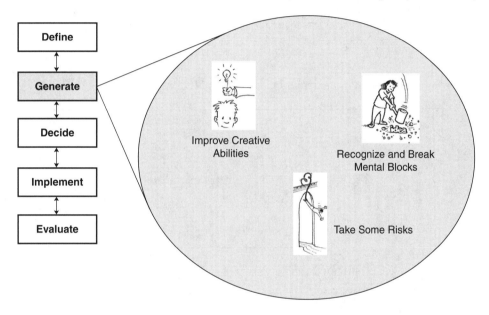

Once you have defined the real problem, you want to make sure you generate the best solution. Sometimes problems may seem unsolvable; at other times, they may appear to have only one less-than-ideal solution. In such a situation, you can use the idea generation techniques described in this chapter to lead you to the best solution. Perseverance is perhaps the most notable characteristic of successful problem solvers, so don't become discouraged when solutions aren't immediately evident. Many times mental blocks may hinder your progress toward a unique solution. The first step in overcoming these blocks is to recognize them. You can then use blockbusting techniques to move forward toward the best solution.

What is the nature of these mental blocks and what causes them? Some common causes of blocks have been summarized by Higgins et al.:[1]

Common Causes of Mental Blocks

Defining the problem too narrowly	Assuming there is only one right answer
Getting "hooked" on the first solution that comes to mind	Trying to get by with a solution that almost works (but really doesn't)
Being distracted by irrelevant information	Being too anxious to finish

There is a direct correlation between the time people spend "playing" with a problem and the diversity of the solutions they are able to generate. Don't be afraid to "play" with the problem by thinking about it in unconventional ways or sleeping on it.

RECOGNIZING MENTAL BLOCKS

Let's look at how easy it is to have a conceptual block to a problem. Try this exercise *before* you read the solutions provided later in the chapter.

Before turning the page, draw four or fewer straight lines (without lifting the pen from the paper) that will pass through all nine dots (Adams, pp. 16–20).[2] Next, draw three straight lines that pass through all the dots.

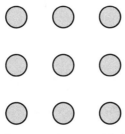

The Nine-Dot Problem

This puzzle is very difficult to solve if you do not cross the imaginary boundary created by the eight outer dots. Another common assumption that is not part of the problem statement is that the lines must go through the centers of the dots. Possible solutions are provided on the next page.

Conceptual blockbusting[2] focuses on the cultivation of idea-generating and problem-solving abilities. The first step to becoming a better problem solver is to understand what conceptual blocks are and how they interfere with problem solving. A **conceptual block** is a mental wall that prevents the problem solver from correctly perceiving a problem or conceiving its solution. The most frequently occurring conceptual blocks are perceptual blocks, emotional blocks, cultural blocks, environmental blocks, intellectual blocks, and expressive blocks.

Perceptual blocks are obstacles that prevent the problem solver from clearly perceiving either the problem itself or the information needed to solve it. A few types of perceptual blocks are described here.

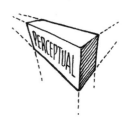

- *Stereotyping.* Survival training teaches individuals to make full use of all the resources at their disposal when they face a life-threatening situation. For example, if you were stranded in the desert after the crash of your small airplane, you would have to make creative use of your available resources to survive and be rescued. Consider the flashlight that was in your toolkit. The stereotypical uses for it would be for signaling, finding things in the dark, and so on. But how about using the batteries to start a fire, the casing for a drinking vessel for water that you find in the desert cacti, or the silver casing reflector as a signaling mirror in the daylight?

- *Limiting the Problem Unnecessarily.* The nine-dot problem is an example of limiting the problem unnecessarily. You must explore and challenge the boundaries of the problem if you hope to find the best solution.

- *Saturation or Information Overload.* Too much information can be nearly as big a problem as not enough information. You can become overloaded with minute details and be unable to sort out the critical aspects of the problem. Air traffic controllers have learned to overcome this kind of perceptual block. They face information overload regularly in the course of their jobs, particularly during bad weather. They are skilled in sorting out the essential information to ensure safe landings and takeoffs for thousands of aircraft daily.

Conceptual Blockbusting:
Two Solutions to the Nine-Dot Problem

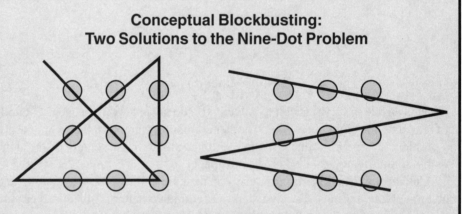

Two Solutions to the Nine Dot Problem

Several other creative solutions to the nine-dot problem exist as well. These include rolling up the piece of paper so that it is cylindrical in shape and then drawing one line around the cylinder that passes through all nine dots, or photo-reducing the nine dots and then using a thick felt pen to connect them with a single line. Another suggestion is to crumple up the piece of paper and stab it with a pencil (this is a statistical approach that may require more than one attempt to hit all the dots).

The purpose of this exercise is to show that putting too many constraints (either consciously or unconsciously) on the problem statement narrows the range of possible solutions. Normally, novice problem solvers will not cross a perceived imaginary limit—a constraint that is formed unconsciously in the mind of the problem solver—even though it is not part of the problem statement. Whenever you are faced with a problem, recall the nine dots to remind yourself to challenge the constraints.

Emotional blocks interfere with your ability to solve problems in many ways. They decrease the amount of freedom with which you explore and manipulate ideas, and they interfere with your ability to conceptualize fluently and flexibly. Emotional blocks also prevent you from communicating your ideas to others in a manner that will gain their approval. Some types of emotional blocks are described here:

- *Fear of Risk Taking.* This block usually stems from childhood. Many people grow up being rewarded for solving problems correctly and punished for solving problems incorrectly. Implementing a creative idea is like taking a risk. You take the risk of making a mistake, looking foolish, losing your job, or (in a student's case) getting an unacceptable grade. Some ideas for overcoming the fear of risk taking will be discussed later in this chapter.

- *Lack of Appetite for Chaos.* Problem solvers must learn to live with confusion. For example, the criteria for the best solution may seem contradictory. You have to be willing to deal with the chaos of not knowing an answer and sorting through details in order to solve a new problem. Not everyone is willing to do that!!

- *Judging While Generating Ideas.* Judging ideas too quickly can discourage even the most creative problem solvers. It is important that a positive creative environment is maintained throughout the brainstorming process so all members are able to participate fully. Wild ideas, although often impractical, can sometimes trigger feasible ideas that lead to innovative solutions; however, these wild ideas are often the ones individuals are discouraged from sharing when ideas are being judged. This block can be avoided by complementing ideas that are truly unique, even if they aren't the perfect solution.

- *Lack of Challenge.* Sometimes problem solvers don't want to get started because they perceive the problem as being too trivial and easily solved. They believe that the problem is not worthy of their efforts.

- *Thinking All or Some Part of the Problem Cannot Be Solved.* Many times this block is related to a lack of energy, or it is used as an excuse to not do work on the problem because "you will never" solve *that part* of the problem. This statement is Fallacy in Logic no. 8, "The Perfect Solution," as described in Chapter 3.

- *Inability to Incubate.* Rushing to solve the problem just to get it off your mind can create mental blocks.

> *Failure is not failure, but an opportunity to begin again, more intelligently.*
> —Henry Ford

Cultural blocks are acquired by exposure to a given set of cultural patterns, whereas environmental blocks are imposed by our immediate social and physical environment. One type of cultural block is the failure to consider an act that causes displeasure or disgust to certain members of society. To illustrate this type of block, Adams cites the following problem:[2]

Being aware of potential conceptual blocks is the first step to overcoming them.

Rescuing a Ping Pong Ball

Two pipes, which serve as pole mounts for a volleyball net, are embedded in the floor of a gymnasium. During a game of ping pong, the ball accidentally rolls into one of the pipes because the pipe cover was not replaced (see below). The inside pipe diameter is 0.06 inch larger than the diameter of the ping pong ball (1.50 inches), which is resting gently at the bottom of the pipe. List as many ways as you can think of (in five minutes) to get the ball out of the pipe without leaving the gym or damaging the ball, pipe, or floor. You are one of a group of six people in the gym, along with the following objects:

A 15-foot extension cord   A bag of potato chips

A file   A chisel

A wire coat hanger   A carpenter's hammer

A monkey wrench   A flashlight

A common solution to this problem is to smash the handle of the hammer with the monkey wrench and to use the splinters to obtain the ball. Another less obvious solution is to urinate in the pipe. Many people do not think of this solution because of a cultural block, because urination is considered a "private" activity in many countries.

Environmental blocks are distractions (phones, interruptions) that inhibit deep, prolonged concentration. Working in an atmosphere that is pleasant and supportive most often increases the productivity of the problem solver. A recent study showed that while working in an area such as Starbucks, people are more creative because there is no one walking into their office to interrupt their thoughts, no landline calls, just white noise that is easily blocked out. Conversely, working under conditions where there is a lack of emotional, physical, economical, or organizational support to bring ideas into action usually has a negative effect on the problem solver and decreases the level of productivity. Ideas for establishing a working environment that enhance creativity were presented in Chapter 2.

Intellectual blocks can occur as a result of inflexible or inadequate uses of problem-solving strategies. A lack of the intellectual skills necessary to solve a problem can certainly be a block, as can a lack of the information necessary to solve the problem. For example, attempting to solve complicated satellite communications problems without sufficient background in the area would soon result in blocked progress. Additional background, training, or resources may be necessary to solve a problem. Don't be afraid to ask for help.

Expressive blocks—that is, the inability to communicate your ideas to others, in either verbal or written form—can also hinder your progress. Anyone who has played a game of charades or Pictionary can certainly relate to the difficulties that this type of block can cause. Make sketches and drawings, and don't be afraid to take the time to explain your problem to others.

As we have just seen, many types and causes of mental blocks exist. If you find your problem-solving efforts afflicted by one of them, what can you do? Try one of the blockbusting techniques found on the Web site. A great way to avoid blocks altogether is to increase your creativity by learning new attitudes, values, and ways of approaching and solving problems and by heeding the guidelines presented in the next section.

IMPROVING YOUR CREATIVE ABILITIES

It is now established that everyone is innately creative and that they can enhance their creativity by practicing regularly. As with any other skill, the more you practice or concentrate on it, the better you get. Raudelsepp[3] has suggested a variety of techniques that can be used to improve your creativity, which you should try to practice as often as you can. These techniques are listed in the following table.

Improving Your Creative Abilities

Keep track of your ideas at all times.	
	Many times ideas come at unexpected times. If an idea is not written down within 24 hours, it will usually be forgotten. Some people even keep a notepad and pencil at their bedside in case they wake up in the night with a creative idea and want to write it down.
Pose new questions to yourself every day.	
	An inquiring mind is a creatively active one that enlarges its area of awareness. If you are doing a homework problem, ask yourself how to make the problem more difficult or more exciting. Apply the critical thinking questions discussed in Chapter 3 to *yourself.*

Continues

Improving Your Creative Abilities (*Continued*)

Learn about things outside your specialty.	
	Use cross-fertilization to bring ideas and concepts from one field or specialty to another. Consider how different classes in school relate to each other or how you can use your abilities in one activity for another purpose. Cross-fertilization is further examined in Chapter 7.
Avoid rigid, set patterns of doing things.	
	Overcome biases and preconceived notions by looking at the problem from a fresh viewpoint, *always* developing at least two or more alternative solutions to your problem.
Be open and receptive to ideas (yours and others).	
	Rarely does an innovative solution or idea arrive complete with all its parts ready to be implemented. New ideas are fragile; keep them from breaking by seizing the tentative, half-formed concepts and possibilities and developing them.
Be alert in your observations.	
	This principle is a key to successfully applying the Kepner–Tregoe approaches discussed in Chapter 8. Keep alert by looking for similarities, differences, and unique and distinguishing features in situations and problems. The larger the number of relationships you can identify, the better your chances of generating original combinations and creative solutions.
Learn to know and understand yourself.	
	Deepen your self-knowledge by learning your strengths, skills, weaknesses, dislikes, biases, expectations, fears, and prejudices.

Keep abreast of your field.	
	Read the magazines, trade journals, and other literature in your field to make sure you are not using yesterday's technology to solve today's problems.

Keep your sense of humor.	
	You are more creative when you are relaxed. Humor aids in putting your problems (and yourself) in perspective. Many times it relieves tension and makes you more relaxed.

Engage in creative hobbies.	
	Hobbies can also help you relax. Working puzzles and playing games both keep your mind active. An active mind is necessary for creative growth.

Have courage and self-confidence.	
	Be a paradigm pioneer. Assume that you can and will solve the problem as described in Chapter 2. Don't be afraid to take a risk. Persist and have the tenacity to overcome obstacles that block the solution pathway.

Adopt a risk-taking attitude.	
	Fear of failure is the major impediment to generating solutions that are risky (i.e., have a small chance of succeeding) but would have a major impact if they are successful. Outlining the ways you could fail and then ways you would deal with these failures will reduce this obstacle to creativity. Some ways you can practice risk taking are challenging established patterns of doing business within your organization, trying a new sport, singing at a karaoke bar, or volunteering to organize a group activity.

RISK TAKING

We just saw in the table "Improving Your Creative Abilities" that risk taking is important for improving your creativity. What are risks? **Risks** are actions, with no certainty of succeeding, which require significant effort, resources, and/or time.

If the actions are successful, however, they will have a major impact. Truly innovative solutions that make a significant difference in your life, organization, or community are almost never found without some risk taking. Of course, the risk taking we encourage you to engage in is not risk taking for the sake of taking a risk. Rather, it is *measured or prudent risk taking*. Know the consequences if your risk fails as well as the consequences of *not* taking the risk by making a list of pros and cons to help decide whether or not to take the risk. Use critical thinking to weigh and assess the consequences of both positive and negative outcomes. Recognize that whenever you take a risk, there will most likely be someone out there to criticize it. Don't be too sensitive to criticism.

Cautions

Do not take risks just for the sake of saying that you are a risk taker! You must have a reason for taking a prudent risk as well as identifying a potential gain, but you also must be willing to live with the consequences if what you hope for doesn't come through. For an example of a risk gone wrong see "The Risk of Risk Taking" on the Professional Reference Shelf on the Web site.

Management Takes a Risk

In 1959, Xerox completed the first prototype of a photocopier and initiated plans to send the new device into mass production. The company soon discovered that mass-producing dependable, high-quality machines would require a major financial investment. Consequently, it hired Arthur D. Little, a consulting firm, to carry out a financial and marketing analysis of the proposed project.

The analysis concluded that no more than 5,000 units of the photocopiers would sell, which might not justify such a large capital investment. The project looked like it would end before it began. Fortunately, Xerox's management took a risk and went ahead with the project.

Epilogue: Ten years later, Xerox passed the billion-dollar mark in sales of copiers as a result of its revolutionary office copiers.

Source: Adapted from Smith, Douglas K., and Robert C. Alexander, *Fumbling the Future*, William Morrow, New York, 1988.

Management Fails to Take a Risk

In 1973, Xerox developed the Alto System, the first personal computer (PC), the first handheld mouse, and the first word processing system. A survey showed there was no market for PCs, however, so the company decided not to take the risk and as a result did not market the Alto.

Epilogue: Xerox was the first firm to develop the PC, yet it remains a "copier company" to many people. By contrast, the revenues from Apple's and IBM's PCs measured in the billions of dollars by 1981.

Source: Adapted from Smith, Douglas K., and Robert C. Alexander, *Fumbling the Future*, William Morrow, New York, 1988. Photo of computer: Fer Gregory/Shutterstock

Coca-Cola's Failure in Indonesia

A major marketing campaign was initiated to introduce Coca-Cola into Indonesia. The campaign cost millions of dollars. Indonesians tried Coca-Cola and did not like it, so sales dropped off rapidly.

Coca-Cola failed! But did the company quit? No! It investigated *why* the sales dropped off. During its research, the company found that Indonesians were used to tea, coffee, and tropical soft drinks and were unaccustomed to carbonated drinks. In response, Coca-Cola phased in carbonation by introducing it into strawberry-, pineapple-, and banana-flavored soft drinks.

Epilogue: The sales of Coca-Cola in Indonesia now surpass the sales of local tropical drinks.

Source: Adapted from Mattimore, Bryan W., *99% Inspiration*, Amacom, USA, 1994, p. 29.

We have just seen three case histories in which companies faced the prospect of taking a risk. In the case of Xerox and the Alto, management chose not to take a risk, which resulted in the loss of billions of dollars in potential revenues. So why don't more companies and individuals take risks? In many cases the reason is fear of failure.

Many people and companies believe that the ideal learning curve is the one shown on the left below. It's a smooth upward journey to success. In reality, the journey to success is more likely to be peppered with setbacks or negative events, as shown in the figure on the right. These are not failures; **they are events on the learning curve**. You should use the knowledge gained from these events constructively so that your chances of success will be even greater on the next try. The only time failure occurs is when there is a negative event and you do not learn anything. To make a breakthrough, you must be willing to make mistakes.

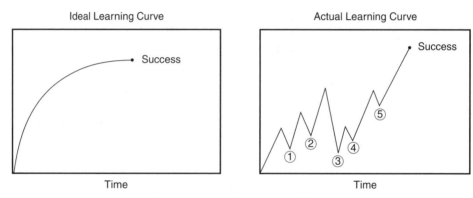

Progress as a Function of Time

In 1958, Tom Monaghan opened a pizza store in Ypsilanti, Michigan. After a year and a half, it went out of business. Was this a failure? It would have been if Tom had not tried to learn the reason his store went belly-up. Instead, Tom researched the reason for the closing and, after some careful planning, opened another store in 1960. This store was the first in what went on to become a worldwide pizza chain, Domino's. To Monaghan, the closing of his first pizza store was not a failure, but rather an event on the learning curve.

In the past, the University of Michigan's business school offered a course called Failure 101. The basic premise of this course is to encourage risk taking by teaching the students not to be afraid to fail with the ideas they have generated.

The course provided many opportunities for students to fail on a number eventually developed into major successful ventures, such as Tom Monaghan's pizza shop. The first car rental agency opened by Warren Avis did not succeed. Glue developed by the 3M Company didn't stick well enough and was nearly abandoned until someone used it to develop Post-it notes. Walt Disney's first animation company went bankrupt, prompting his move to California and the creation of his new studio where Mickey Mouse was launched. When the Petrossian brothers, who fled from Russia in 1917, introduced caviar at the Ritz Hotel in Paris, the French made ready use of the nearby spittoons. The brothers were quite discouraged and might easily have given up on the idea. Fortunately, they persisted and overcame this first rejection, and today Petrossian Caviar is sold throughout France at prices of as much as $1000 per pound. J. K. Rowling's first Harry Potter book was rejected by 12 publishers. Other examples include Michael Jordan being cut from his high school basketball team, John Grisham having his first two books turned down, and Oprah Winfrey not succeeding as a news reporter.

The instructor of the Failure 101 course also gave the students projects that at first seemed nearly impossible.[4] Consider the following projects:

- Design and prepare a marketing campaign for reversible baby diapers.
- Plan a marketing campaign to open a maternity shop in a retirement center.

The students enthusiastically tackled these opportunities to fail. For example, they suggested manufacturing baby diapers that are blue on one side and pink on the other side and encouraging grandparents-to-be in the retirement center to have their expectant daughters come and spend quality time shopping together.

If major breakthroughs are to be made, risks must be taken. Failures resulting from these risks will occur but should not deter future risk taking. The knowledge gained from these failures should be used constructively so that the chances of success will be even greater on the next try.[4] The following table shows what we should do to help overcome the fear of failure.

Overcoming Fear of Failure

1. Outline what the risk is and explain why the risk is important for you to take.

2. Describe the worst possible outcome if you take the risk and fail.

3. Describe your options given the worst possible outcome.

4. Describe what you could learn from the worst possible outcome.

As we have just seen, risk taking often leads to major breakthroughs. The question for you is this: "Will you be able to have or develop a mindset that will allow you to take risks, or will you be like the manager at Xerox who did not market the Alto?" To help develop a risk-taking mindset, practice taking small risks whenever possible, such as volunteering to speak at a conference or meeting, challenging established patterns of doing business in your organization, or trying a new sport. Let's examine a real-life example of a situation where a risk was not taken but maybe would have been beneficial.

Overcoming Fear of Failure: Nokia's Bad Call on Smartphones

In the late 1990s, Nokia Corporation presented wireless carriers and investors with two new devices from their research and development department. The first, a touch-screen phone with a single button that could locate eateries and play games, was strikingly similar to the device now known as the iPhone, which would be released seven years later. The second was a tablet computer similar to the iPad.

Neither device ever made it to consumers, as Nokia was stuck in paradigm paralysis, that is, being complacent with current business and products, saying what we have is enough. Nokia was, at the time, a leader in the lucrative "dumb" phone market and its leadership team was apparently not willing to be paradigm pioneers in recognizing these big ideas and taking the risk of moving them forward to fruition.

Epilogue: Apple later shifted the paradigm and brought its devices, the iPhone and the iPad, to market in the 2000s and became one of the most successful companies of the new millennium.

Let's examine what could have happened if Nokia's leadership team had applied the four steps of overcoming the fear of failure to the situation.

Step 1. Outline what the risk is and explain why the risk is important for you to take.

The risk in this situation is for Nokia to invest money and resources to bring the new products to market. It is unsure whether the new products will be successful and would cause the company to lose time and money. The company is in a safe place being a leader in the "dumb" phone market, and departing from its current position by heavily investing in a revolutionary phone could jeopardize its profits.

The risk is important to take because the products the Nokia team has developed have the potential to change the cellular device market by offering customers new features and capabilities, which could create a large demand for their products. In doing so, the company would make a huge return on its investment. Complacency with the current products is not a reason to pass over evaluating taking the risk of bringing new products to market.

Step 2. Describe the worst possible outcome if you take the risk and fail.

The company invests in the new products and no one is interested in buying these new devices: in other words, they fail. The company loses significant amounts of money on the product development and has spent significant time and money on this project while they could have been working on other projects. The company releases two bad products to the market giving customers a bad opinion of the company and providing the opportunity for other competitors to gain insights into Nokia's newest technologies and learning from their mistakes to gain success.

Step 3. Describe your options when given the worst possible outcome.

After "failure," the company would be able to abandon the new projects in favor of returning its focus to "dumb" phones or continue to invest in the "failed" projects based on feedback from customers and try to make them into something that customers will want to buy.

Step 4. Describe what you could learn from the worst possible outcome.

By "failing," the company could learn what the customers do not want at that time by producing new products that no one wants to buy. Devices with no demand can provide insight as the customer likely wants some of the opposite features, or at least some different combination of features. The company could incorporate these insights into either refining its failed products or into enhancing the "dumb" phones, possibly creating new demands.

Source: "Nokia's Bad Call on Smartphones," *Wall Street Journal*, July 18, 2012.

After thinking through the four steps, it is evident that even if the new products "failed," Nokia could have learned a lot by investing in them and bringing them to market. In hindsight, it is important that the company learn from the missed opportunity and work to better evaluate risks in the future. Not all risks are worth taking, however; the prudent use of risk taking can be a powerful tool for successful problem solving. In Chapter 8, we will discuss a technique called potential problem analysis that can be used to identify the possible downside to problem solutions that may involve risk.

SUMMARY

This chapter presented techniques to help you overcome mental blocks, improve your creative abilities, and adopt an attitude of being willing to take risks. Here are some ways to help you develop more creative solutions:

- Be able to recognize the different mental blocks when they appear (i.e., perceptual, emotional, cultural, environmental, intellectual, and expressive blocks).
- Practice the 12 ways to increase your creative ability.
- Develop a more positive attitude toward risk taking. Before taking on risk, however, make sure you have identified all potential negative outcomes and would be willing to live with those consequences.
- Recognize that "failure" is simply an event on the learning curve *unless* you do not learn from the event—*then* it is a failure.

WEB-SITE MATERIAL (WWW.UMICH.EDU/~SCPS)

- **Learning Resources**
  Summary Notes

- **Professional Reference Shelf**
  1. Goman's Blockbusters
  2. Risk Taking
     - Arsenio Hall
     - The Jolly Green Giant
     - Why Is the Champagne Dry, Charles?
     - The Risk of Risk Taking

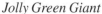

Jolly Green Giant   *Why Is the Champagne Dry, Charles?*

© General Mills, Inc. Reprinted by permission.

REFERENCES

1. Higgins, J. S., et al., "Identifying and Solving Problems in Engineering Design," *Studies in Higher Education*, 14, 2, p. 169, 1989.
2. Adams, James L., *Conceptual Blockbusting: A Guide to Better Ideas*, 3rd ed. © 1986. Reprinted by permission of Da Capo Press, a member of Perseus Books Group.
3. Raudelsepp, E., "Taking This Test to Measure Your Creativity," *Chemical Engineering*, 85, p. 95, July 2, 1979.
4. Matsen, J., *How to Fail Successfully*, Dynamo Publishing, Houston, TX, 1990.

EXERCISES

6.1. Make a list of the following:

    A. The five most important things you learned from this chapter.

    B. The five most interesting things you learned that you could use to make a conversation more interesting.

    C. The five best ideas you generated in the past three days.

    D. The five problems you've recently encountered that were not worth solving.

Mental Blocks

6.2. A. Make a list of the worst business ideas you can think of (e.g., a maternity shop in a retirement village, a solar-powered night-light, reversible diapers).

    B. Take the list you generated in part (A) and turn it around to make the ideas viable concepts for entrepreneurial ventures (e.g., reversible diapers—blue on one side and pink on the other; solar powered night-light—charge in the day to use at night).

6.3. A. Rearrange four pencils to make six equal triangles.

    B. Remove six pencils to leave two perfect squares and no odd pencils.

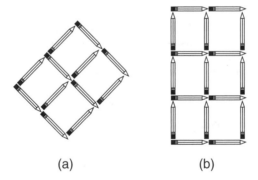

          (a)   (b)

6.4. A. Rearrange three balls so that the triangle points up instead of down.

    B. Moving one black poker chip only, make two rows of four.

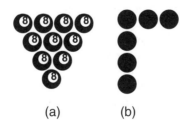

        (a)   (b)

6.5. Fifty-seven sticks are laid out to form the equation. Remove eight sticks to make the answer correct. Do not disturb any sticks other than the eight to be removed. First list any perceived constraints that you initially thought could be blocks to solving this problem.

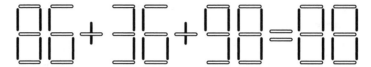

Source: Carter, Phillip J., and Ken A. Russel, *Brain Busters*, Sterling Publishing, New York, 1992.

6.6. A prize is hanging by a string from a 10-foot ceiling. You are seated in an immovable chair six feet away. In your possession are 10 pieces of paper, a pair of scissors, a reel of tape, paper clips, a box of matches, and a ball of string. Suggest ways of obtaining the prize while remaining seated.

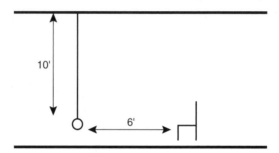

6.7. Identify the mental blocks you encountered in Exercises 6.3 through 6.6.

6.8. Give a specific example for each of the following:

A. Each of Adams's perceptual blocks

B. Each of Adams's emotional blocks

C. Each of Higgins's common causes of mental blocks

Improving Your Creative Ability

6.9. Pick three of the 12 techniques listed in the table on improving your creative abilities to work on in the next month. Outline the first steps on how you will do this.

6.10. Which of the 12 techniques on improving your creative abilities do you believe are your current strengths? Your current weaknesses?

6.11. Create a scenario on how Coca-Cola or Xerox might have used techniques from the table on improving your creative abilities to develop their products.

Risk Taking

6.12. Identify a small, medium, and large risk you should take in the not-too-distant future. Describe your fear of failure for each risk. Apply each of the items in the table on overcoming the fear of failure.

6.13. Develop a scenario for what Xerox's Alto System might have looked like if the company had decided to use the items in the table on overcoming fear of failure when deciding on this product's future.

6.14. What lesson can you take away from the example of the Nokia smartphone and tablet?

6.15. Describe some creative, but prudent, risks that you can take the next time that you search for a new job.

6.16. Discuss some of the points you would consider as you decide whether a risk you are contemplating is appropriate.

6.17. Think of a situation where you took a risk. Describe the fears that you had to overcome to be able to take the risk. What were the negative consequences that you had to consider might occur as a result of your decision to proceed? How would you have dealt with those consequences if the risk had not been successful?

6.18. Just as Nokia made a "bad call" on smartphones, the imaging company Kodak made a bad call on the digital camera. Kodak refused to move quickly into the digital camera business and instead stuck with the traditional film market despite inventing key digital camera technology. This decision to not move forward or take the risk of entering the digital market early eventually landed the company in bankruptcy in 2012. Apply the four steps of overcoming the fear of failure to Kodak's opportunity to take a risk and move into the digital camera market.

FURTHER READING

Adams, James L. *Conceptual Blockbusting: A Guide to Better Ideas*, 3rd ed. Addison-Wesley, Stanford, CA, 1986.

Von Oech, Roger. *A Whack on the Side of the Head: How You Can Be More Creative*, revised edition. Warner Books, New York, 1990.

7   GENERATING SOLUTIONS

Nothing is more dangerous than an idea, when it is the only one you have.

—Emile Chartier

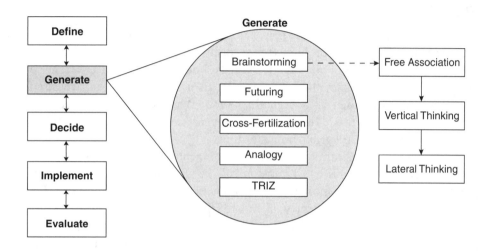

This chapter describes a variety of idea generation techniques that will enhance your ability to *think out of the box* and develop solutions that are truly innovative. These techniques, shown in the diagram above, should be practiced at every opportunity so that they become second nature to you when working through the problem-solving algorithm.

r69photo/Shutterstock

"The only ones who should not think out of the box are cats."

BRAINSTORMING

Brainstorming—one of the oldest techniques to stimulate creativity—is a familiar and effective technique for generating solutions. It provides an excellent means of getting the creative juices flowing. Recent surveys of people working in industry show that brainstorming is routinely used as an effective tool, not only for two or three individuals discussing a problem in an informal setting, but also in more formal, large-group problem-solving sessions.

We begin our process with free association—that is, by writing down as many suggestions as we can without judgment of the feasibility. At first the flow of ideas will be very rapid; however, after a while you will observe that the rate at which new ideas or suggestions are produced becomes quite slow. At this point we need to use triggers to rejuvenate the rate of suggestions. Some of the most commonly used triggers are vertical thinking, lateral thinking, TRIZ, cross-fertilization, and futuring. We begin with free association. It is followed by vertical thinking using SCAMPER, which reviews and builds on and expands the initial list of ideas.[1] We then move to lateral thinking using random stimulation and other people's views. Finally, we engage in futuring, cross-fertilization and analogy, and TRIZ. After we have used these triggers to generate as many ideas as possible we organize our ideas in a fishbone diagram.

FREE ASSOCIATION

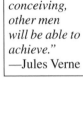

"What one man is capable of conceiving, other men will be able to achieve."
—Jules Verne

Typically, the initial stages of idea generation begin with an unstructured free association of ideas to solve the problem (brainstorming). During this activity, the group creates lists of all possible solutions. These lists should include wild and crazy solutions or unusual solutions without any regard to their feasibility because these solutions could spark an idea for a logical solution. When brainstorming in groups, people can build upon one another's ideas or suggestions. This triggering of ideas in others is key to successful group brainstorming.

You can use brainstorming to improve your creativity in technical areas. When you finish a homework problem, brainstorm all the ways you could have worked the problem incorrectly, with more difficulty, more easily, or in a more exciting way. Brainstorm a list of all the things you learned from the problem or ways you could extend the problem. Continually ask "What if?" questions. For example, what if someone suggested doubling the size of the equipment to double production? Brainstorm all the advantages and disadvantages of making such a change.

Another critical component of group brainstorming is maintaining a positive group attitude. No negative comments or judgments are allowed during this stage of the solution process: Reserve your evaluation and judgment until later. As more ideas are generated, the group stands a better chance of devising an innovative, workable solution to the problem at hand. Nothing kills a brainstorming session faster than negative comments. If negative comments are not kept in check by the group leader, the session will usually be reduced to one of "braindrizzling."

Comments That Reduce Brainstorming to Braindrizzling

- That won't work.
- It's against our policy.
- It's not our job.
- We haven't done it that way before.
- We don't have enough time.

- That's too expensive.
- That's too much hassle.
- That's not practical.
- That's too radical.
- We can't solve this problem.

We have conducted numerous brainstorming exercises with groups of students. An example of an unstructured session is shown on the next page.

As mentioned earlier, typically the ideas flow quickly at first and then slow abruptly after several minutes. That is, the process hits a "roadblock." These roadblocks hinder our progress toward a solution. Luckily, we can use some blockbusting techniques to help overcome these mental blocks and generate additional alternatives.

Brainstorming Activity

Problem Statement

Suggest uses of old cars as equipment for a children's playground.

Ideas Generated by Free Association

- Take the tires off and roll them along the ground.
- Get on the roof and use the car as a slide.
- Take the seats out and use them as a bed on which to rest between activities.
- Teenagers could take the engine apart and put it back together.
- Cut the car apart and turn it into a 3-D puzzle.
- Make a garden by planting flowers inside.
- Use the tires to crawl through as an obstacle course.
- Make the car into a sculpture.
- Take the doors off and use them as goals for hockey.

VERTICAL THINKING

Vertical thinking can build on the ideas already generated (piggybacking) or it can look at the different parts of the problem in an effort to generate new ideas. One of the vertical thinking techniques is SCAMPER, an acronym for a useful list of active verbs that can be applied as stimuli to make you think differently about a problem. SCAMPER was defined by Robert Eberle, and it is a modification of the work known as Osborn's checklist.

SCAMPER

Substitute:	Who else, where else, or what else could be substituted for? Substitute another ingredient, material, or approach?
Combine:	Combine parts, units, ideas? Blend? Compromise? Combine from different categories?
Adapt:	How can this (product, idea, plan, etc.) be used as is? What are other purposes it could be adapted to?
Modify:	Magnify? Minify? Change the meaning, material, size, etc.?
Put to other use:	How can you put the thing to different or other uses?
Eliminate:	Remove something? Eliminate waste? Reduce something?
Rearrange:	Interchange components? Change pattern, pace, schedule, or layout?

Magnify

Minify

Rerraange

Com⟶ ⟵ bine

Example of Vertical Thinking Using SCAMPER

Continuing with the playground equipment example …

Substitute: Use the cars' seats in swings.

Combine: Use the side panels or roof of the car to make a huge canopy or fort.

Adapt: Take the hood off the car, and use it as a toboggan in winter.

Modify: Crush the cars into cubes, and allow the kids to climb on the blocks.

Put to another use: Over-inflate the inner tubes from the tires, and use them to create a "romper room"/jumping pit.

Eliminate: Remove the engines and side panels, and make go-carts.

Rearrange: Turn the car upside down, and use it as a teeter-totter.

77 Cards: Design Heuristics

A new technique that is more extensive than SCAMPER is *design heuristics*.[2,3] This vertical thinking technique includes lists of prompts intended to help designers move through a "space" of possible solutions and also to support designers in becoming "unstuck" when they are struggling to generate more, and different, ideas.[4,5] The 77 design heuristics below are a result of combined outcomes from a designer case study, extractions of characteristics of award-winning products, and protocol studies of designers and engineers of varying expertise levels.

1 Add features from nature
2 Add gradations
3 Add motion
4 Add to existing product
5 Adjust function through movement
6 Adjust functions for specific users
7 Align components around center
8 Allow user to assemble
9 Allow user to customize
10 Allow user to reconfigure
11 Animate
12 Apply existing mechanism in new way
13 Attach independent functional components
14 Attach product to user
15 Bend
16 Build user community
17 Change contact surface
18 Change direction of access

19 Change flexibility
20 Change geometry
21 Compartmentalize
22 Convert 2-D to 3-D
23 Convert for second function
24 Cover or remove joints
25 Cover or wrap
26 Create system
27 Distinguish functions visually
28 Divide continuous surface
29 Elevate or lower
30 Expand or collapse
31 Expose interior
32 Extend surface
33 Extrude
34 Flatten
35 Fold
36 Hollow out
37 Impose hierarchy on functions
38 Incorporate environment
39 Incorporate user input
40 Layer

41 Make components multifunctional
42 Make components attachable or detachable
43 Make product reusable or recyclable
44 Merge functions with same energy source
45 Merge surfaces
46 Mirror or Array
47 Nest
48 Offer optional components
49 Provide sensory feedback
50 Reconfigure
51 Recycle to manufacturer
52 Reduce material
53 Reorient
54 Repeat
55 Repurpose packaging
56 Reverse direction or change angle
57 Roll
58 Rotate

59 Scale up or down
60 Separate parts
61 Slide components
62 Stack
63 Substitute
64 Synthesize functions
65 Telescope
66 Texturize
67 Twist
68 Unify
69 Use alternative energy source
70 Use common base to hold components
71 Use continuous material
72 Use human-generated power
73 Use multiple components for one function
74 Use packaging as functional component
75 Use recycled or recyclable materials
76 Utilize inner space
77 Utilize opposite surface

The design heuristics, which can be used for vertical thinking, are represented on cards that can be found at www.designheuristics.com. Each card includes a description of the heuristic, an abstract image depicting the application of the heuristic, and two product examples that show how the heuristic is evident in existing consumer products. An example card for heuristic 77, "Utilize opposite surface," is below.

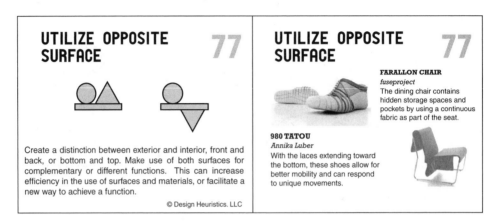

For example, we could use ideas from the 77 cards to develop conceptual designs for a device that utilizes sunlight *to heat and cook food.*[4] Let's use the design heuristics to develop three unique ideas.

In the first idea we combine the strategies of two separate cards—heuristic card number 67, "Twist," and card number 22, "Convert 2-D material to a 3-D object"—to generate a single idea. By combining these cards, we are able to create a spiral-shaped reflector out of a single sheet of metal, capable of concentrating a large amount of light onto a small cooking surface.

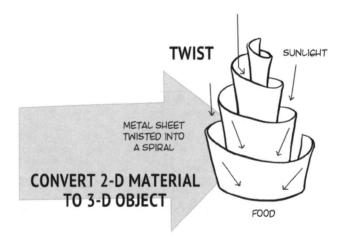

In the second idea, we begin with a simple box with flaps to reflect light into the center, and then use heuristic card 1 and card 23 to transform the ideas. "Mimic natural mechanisms" prompted thinking about how flowers bloom, "Convert for second function" prompted thinking about how the device could function in two different states (closed or open), and "Utilize opposite surface" prompted ideas about mirrors that could be used on the inside to direct light to heat food and about solar panels that could be used on the outside to capture energy and generate heat for the food.

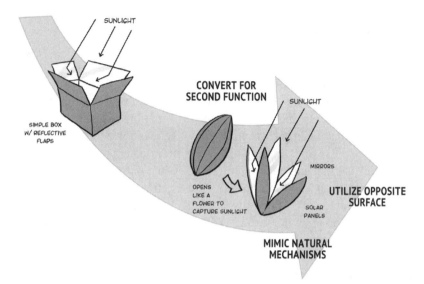

In the third idea we start with heuristic card 15 ("Bend"), 35 ("Fold"), and 65 ("Telescope") to generate an idea for a deployable parabolic reflector. Next, we modified that idea by changing the reflective parabola to be constructed from multiple small pieces of recycled mirrors by using heuristic card 54 ("Repeat") and 75 ("Use recycled or recyclable materials").

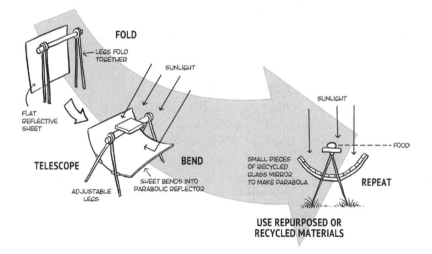

LATERAL THINKING

Stuck

Unstuck

Unlike vertical thinking, which builds on the preexisting ideas, lateral thinking injects ideas that are not related to previous ideas. Edward de Bono, regarded by many as the father of lateral thinking, developed a number of lateral thinking techniques[6] that provide different ways to come at the problem from a new direction and get "unstuck" when you have trouble generating new ideas or solutions. After only one or two times using lateral thinking techniques in brainstorming, you will be convinced that the solutions or ideas you generated are ones that never would have been generated by free association or vertical thinking.

Although one of the first steps in the problem-solving process recommended by experienced engineers is the gathering of information, de Bono cautions problem solvers in this regard. For example, when you begin working on a new problem or research topic, it is normal to read all the information available on the problem. Failing to do so may mean "reinventing the wheel" and wasting much time. However, during the course of gathering this information, you may destroy your chances of obtaining an original and creative solution if you are not careful.

As you read, you will be exposed to the existing assumptions and prejudices that have been developed by previous workers or researchers. Try to remain objective and original, or your innocence will have been lost. De Bono recommends reading just enough to familiarize yourself with the problem and get a "feel" for it. Can you get a "ballpark" answer with a Fermi calculation? At this point you may wish to stop and organize some of your own ideas before proceeding with an exhaustive review of the literature. This strategy allows you to preserve your opportunities for creativity and innovation.

Have you ever heard the old saying, "If it ain't broke, don't fix it"? De Bono claims the attitude reflected by this statement was largely responsible for the decline of U.S. industry in the past few decades. American managers operated in a strictly reactive mode, merely responding to problems as they arose. Meanwhile, their Japanese counterparts were fixing and improving things that weren't problems. Soon, the American "problem fixers" were left behind. To survive in today's business culture, proactive thinking—as opposed to reactive thinking—is required. This shift in thinking patterns requires creativity.

Lateral Thinking Using Random Stimulation

Random stimulation is a technique that is especially useful if we are stuck or in a rut.[6,7] It is a way of generating totally different ideas than previously considered. As a result, it can "jump-start" the idea generation process and free it from whatever current rut it may be in. The authors of the text have taught over 25 short courses in industry and academia and have used a brainstorming exercise with random stimulation. *There has not been **one** occasion* in which one or more of the participants did not say, "You know I would never have thought of that idea had I not used random stimulation."

The introduction of strange or "weird" ideas during brainstorming should not be shunned but rather encouraged. Random stimulation makes use of a random piece of information (perhaps a word culled from the dictionary or a book [e.g., the eighth word down on page 125] or one of the words in the sample list picked by a random finger placement). This word serves as a trigger or switch to change the patterns of thought when a mental roadblock occurs. The random word can be used to generate other words that can stimulate the flow of ideas.

A Short List of Random Stimulation Words

all, albatross, airplane, air, animals, bag, basketball, bean, bee, bear, bump, bed, car, cannon, cap, control, cape, custard pie, dawn, deer, defense, dig, dive, dump, dumpster, ear, eavesdrop, evolution, eve, fawn, fix, find, fungus, food, ghost, graph, gulp, gum, hot, halo, hope, hammer, humbug, head, high, ice, icon, ill, jealous, jump, jig, jive, jinx, key, knife, kitchen, lump, lie, loan, live, Latvia, man, mop, market, make, maim, mane, notice, needle, new, next, nice, open, Oscar, opera, office, pen, powder, pump, Plato, pigeons, pocket, quick, quack, quiet, rage, rash, run, rigid, radar, Scrooge, stop, stove, save, saloon, sandwich, ski, simple, safe, sauce, sand, sphere, tea, time, ticket, treadmill, up, uneven, upside-down, vice, victor, vindicate, volume, violin, voice, wreak, witch, wide, wedge, X-ray, yearn, year, yazzle, zone, zoo, zip, zap

In using the random simulation technique, we randomly put our finger on one of the words in this short list. Suppose it fell upon the word "document." This word makes us think of the word "paper," which makes us think of "art" and continues in the progression shown below. If you get up to 8–10 words and have not come up with a new idea related to the topic you are brainstorming, choose a new word and continue.

Example of Random Stimulation

Problem

Continuing the playground equipment example

Random Simulation Paper Trail

Document → Paper → Art → Colors → Paint → Allow kids to paint graffiti on cars.

This kind of pattern change allows us to view the problem from new perspectives that we had not previously considered.

Lateral Thinking Using Other Points of View

When approaching a problem that involves the thoughts and feelings of others, a useful thinking tool is Other Points of View (OPV).[6] The inability to see the problem from various viewpoints can be quite limiting. And seeing is just the beginning: viewpoints can contain sounds, smells, emotions, and more. Imagining yourself in the role of the other person, or even an inanimate object, allows you to see complications of the problem that you had not considered previously. For example, automotive engineers must be aware of many perspectives if they hope to design a successful vehicle. In particular, they must consider the views of the consumers, their company's marketing personnel, management, the safety department, the financial people, and the service personnel. Failure to take any of these groups' views into account could result in a failed product.

Often the people creating the product or solution are not the ones who will end up using it. This creates issues when the users and creators have fundamentally different knowledge bases or skill sets. This "block" can be overcome by having the creators place themselves in the shoes of the users or even going to the users themselves and finding out how they will use the product or solution. Menlo Innovations, a software development company, employs full-time "high-tech anthropologists" whose jobs are to view things from their clients' perspective. They then make sure Menlo's software engineers create products that are right for the way clients will use them, not the way the engineers think they will be used.

Consider an argument between a newly hired store manager and an employee. The issue at hand is the employee's desire to take a two-week vacation during the store's busiest period, the Christmas season. The manager's main concern is having enough help to handle the sales volume. The employee, however, has made reservations for an Antarctic cruise, one year in advance (with the former manager's approval), and stands to lose a lot of money if he has to cancel them. This problem does not have a solution yet, but by using OPV each person can see what the other person stands to gain or lose from the vacation, and each will have a better understanding of the types of compromises the other person might be willing to make. An example using this technique is shown below.

Example of Other Points of View

Problem

Continuing the playground example

Think about viewing the car from a child's viewpoint. Think about walking around on your knees. How would this change your perspective? That is, imagine the playground from a child's height. What was your favorite toy? How could this be mimicked with used auto parts?

Example

From a child's point of view, the intact car would be an exciting chance to pretend to be a grownup. Just take off the door, remove other dangerous equipment, and let the kids pretend to drive. Just leave the car as it is. Horns can be loud and obnoxious to adults, but children might enjoy being able to create sound, so consider leaving the horn in.

Example

From the bird's point of view, a bird bath would be nice: Fill the hub caps with water to attract birds to sit in for the children to enjoy.

ORGANIZING BRAINSTORMING IDEAS: THE FISHBONE DIAGRAM

Fishbone diagrams are a graphical way to organize and record brainstorming ideas. Such a diagram looks like a fish skeleton (hence the name). To construct a fishbone diagram, we follow this procedure:

1. Write the real problem you want to solve by generating ideas in a box or a circle to the right of the diagram. Next, draw a horizontal line (the backbone) extending from the left side to the box:

Solutions to the Real Problem

2. Categorize the potential solutions into several major categories (e.g., whole car, parts, painting) and list them along the bottom or top of the diagram. Extend diagonal lines from the major categories to the backbone. These lines form the basic skeleton of the fishbone diagram.
3. Place the potential solutions related to each of the major categories along the appropriate line (or bone) in the diagram.

A fishbone diagram for organizing the ideas for the cars as playground equipment problem is shown on page 159. The most difficult task in constructing such a diagram is deciding which major categories to use for organizing the options. In this example, we have selected "Painting," "Whole Car," and "Parts." The ideas that were generated fall neatly into these categories. Other categories often used in fish-bone diagrams include personnel, equipment, method, materials, and the environment. This activity of sorting and organizing the information is a very valuable component of the solution process.

By reviewing the fishbone diagram, we can evaluate the solutions that have been generated. We have put a structure to the solutions, organizing them and allowing us to "attack" the problem from a number of different fronts if we choose. Clearly, fishbone diagrams can be very helpful in visualizing all the ideas that you have generated.

Painting
- Let kids paint graffiti on the cars.
- Paint targets and let kids throw balls at them.
- Paint the car as a covered wagon and let the kids pretend to be cowboys.

Whole Car
- Turn the car into a teeter-totter (upside down).
- Turn the car into a go-cart.
- Crush the car and make blocks from it.
- Let kids drive the car as is.
- Open the car's doors and use them as goals for field hockey.

Parts
- Use the seats in swings.
- Use the roof and doors as part of a fort.
- Use the tires' inner tubes as part of an obstacle course (to jump on).
- Use the car's hood as a toboggan.
- Use the car's springs for a wobble ride.

We now choose the best idea to put on each of the major bones of the fish:
Painting → Graffiti
Parts → Tire Inner Tubes
Whole Car → As Is
We could go even further and choose the best of the major branches:
Whole Car → As Is

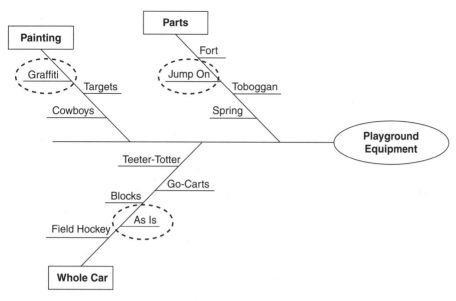

Fishbone Diagram

CAUTIONS

Another step, which has been omitted in this example, is actually necessary before we choose the solutions to put on the fishbone diagram. That step is evaluating the suggestions generated. In the preceding example, we should have addressed a pressing issue—namely, the children's safety—for each of the ideas generated. We could address this safety issue by using the Kepner–Tregoe approach of *potential problem analysis,* which is discussed in Chapter 8.

BRAINWRITING

If you have no one to interact with, or prefer working alone, you can use another technique to generate ideas: brainwriting. In brainwriting you follow the same procedure as in brainstorming (e.g., free association, SCAMPER, random stimulation, futuring) and write down your ideas as quickly as you generate them, without pausing or stopping to evaluate the ideas. Also, keep a notebook handy to write down ideas whenever they occur to you because they often come at unexpected times. After you have completed your list, organize your ideas (solutions) into a fishbone diagram. In fact, recent research has shown that some types of individuals generate better ideas using brainwriting than in group brainstorming.

FUTURING

Futuring is a blockbusting technique that focuses on generating solutions that may not be technically feasible today but might become practical in the future. In futuring, we ask questions such as "What are the characteristics of an ideal solution?"

and "What currently existing problem would make our jobs easier when solved, or would solve many subsequent problems, or would make a major difference in the way we do business?" One of futurist Joel Barker's key ideas is that you should be bold enough to suggest alternatives that promise major advances, yet may have only a small probability of success.

The rules for futuring are relatively simple: Try to imagine the ideal solution without regard to whether it is technically feasible. Then begin by making statements such as "If [this] happened, it would completely change the way I do business." For example, the University of Michigan's College of Engineering Commission on Undergraduate Education used futuring exercises to help formulate the goals and directions of its engineering education program for the twenty-first century. The members of the commission were asked, "What do you see the student doing in 2020?" Their answers included these responses:

- "I see students using interactive computing to learn all their lessons. There are animations and simulations of processes where the students can change operating parameters and get instant visual feedback on their effect."
- "I see lecture halls where the lecturer is a hologram of the most authoritative and dynamic professor in the world on that particular topic."

In futuring, you visualize the idealized situation that you would like to have and then work on devising ways to attain it. Here's the futuring process in a nutshell:

The Futuring Process

1. Examine the problem carefully to make sure the real problem has been defined.
2. Imagine yourself at some point in the future after the problem has been solved. What are the benefits of having a solution?
3. "Look around" in the future. Try to imagine an ideal solution to the problem at hand without regard to its technical feasibility. Remember, in the future, anything is possible.
4. Make statements such as "If only [this] would happen, I could solve [this problem]."
5. Dare to change the rules! The best solutions to some problems are contrary to conventional wisdom.

Futuring in Action: Highway Congestion No More

Highway congestion in the areas of San Francisco and Los Angeles is estimated to cost over $100 billion annually in wasted time and fuel. One proposed solution is to build a bullet train between the two cities. Another is to build new highways or widen current ones. All of these options require large investments in infrastructure and will require even more money in large annual maintenance costs to keep them running. Let's try an exercise in futuring to find a better solution.

Let's imagine ourselves in the *future* with a transportation system that allows all travelers to get where they want to be efficiently and safely. How will this transportation system work? It will be highly successful because the cars are not driven by people but by computers that allow for maximized efficiency by zipping passengers from point A to point B quickly and safely by communicating with other vehicles to compute the optimal route of travel. How can aspects of this future solution be brought to the present?

"*What if*" we put computers in cars that would interact with the traffic lights, sensors, and road signs? Current traffic light systems are based on largely out-of-date historical data, wasting millions of hours of travel time each year. Combined with information from traffic sensors in all locations of the city, and digital road signs and arrows to give information to drivers, computerized traffic lights could route traffic around congestion or even re-route traffic to avoid congestion all together. A computerized system, which is currently possible, would cost a fraction of building new highways or a bullet train and be much easier to maintain and adapt as travel patterns change over the years.

Continues

Summary

Define the Real Problem: The problem is not how to deal with congestion but rather how to move people efficiently.

Imagine the Future: Congestion is a thing of the past as computer-controlled cars can move quickly and safely along the most efficient route possible at any given time.

Generate Solutions: Instead of making all cars have computers, which is technically feasible but not economical, combine traffic sensors with computers in traffic lights to route traffic in more efficient ways around cities.

Source: Adapted from "Opinion: Paving The Way for Driverless Cars," Clifford Winston, *Wall Street Journal*, July 18, 2012.

CROSS-FERTILIZATION

It is well documented that a number of the most important advances in science, engineering, art, and business have come from cross-fertilization and analogies with other disciplines. In this process, ideas, rules, laws, facts, and conventions from one discipline are transferred to another discipline.

Cross-fertilization utilizes unique knowledge and skill sets of individuals and groups with different backgrounds by applying expertise in new disciplines. A major chemical company arranged a series of lunches where accountants would sit with chemists, mathematicians with salespeople, engineers with advertising people, biologists with human resources, and so on. The idea was to learn what ideas, heuristics, and paradigms might be brought from one discipline to another.

To practice generating ideas by cross-fertilization, you might ask what each of the following pairs would learn if they went to lunch or dinner together that would improve themselves and/or the way they perform their jobs:

A beautician and a college professor
A police officer and a software programmer
An automobile mechanic and an insurance salesperson
A banker and a gardener
A choreographer and an air traffic controller
A maitre d' and a pastor

Many new technologies, features, and business have been created from ideas conceived using cross-fertilization. One example is Virgin Atlantic Airlines.

Virgin Atlantic Airlines: Entertainment on the Go

In the 1980s media mogul Sir Richard Branson, founder and owner of the Virgin Records and Virgin CD and Video outlet stores, joined forces with airline industry veterans to start a new airline, Virgin Atlantic Airlines, which became a major airline in the United Kingdom. The business venture used cross-fertilization between the airline industry experts' knowledge of how to operate an airline with Branson's entertainment industry understanding of customer enjoyment. The result was a well-operated airline that brought customers a unique and enjoyable in-flight and in-airport experience unparalleled in the travel industry because it had entertainment industry flair. Virgin Atlantic leads the industry in providing the most extensive in-flight entertainment options and services, chic airport clubhouses with innovative services like touchdown revival treatments after long flights, and even a wide array of mobile apps to help travelers before, during, and after their trips.

Source: http://www.virgin-atlantic.com/gb/en.html.

The cross-fertilization of ideas from one group to another is a powerful method for adapting ideas from one discipline or profession to solve problems in another discipline or profession. Many times managers will bring together a small group of people from diverse (ethnic, cultural) backgrounds to interact and look at a problem and solution from many vantage points. In Steven Covey's *The 3rd Alternative: Solving Life's Most Difficult Problems*[8] the author presents an example where cross-fertilization is used to create a win-win solution.

Cross-Fertilization in Practice: Eradicating Malaria

DDT was once used successfully in the fight against the deadly disease malaria. Malaria is spread by the bite of a mosquito and when DDT was used to kill mosquitoes, deaths from malaria plunged. However, DDT was banned when environmentalists revealed pesticides were causing significant long-term damage to the environment. Without DDT to kill mosquitoes, there was an instant and distressing increase in the number of malaria deaths. A foundation quickly brought together a diverse group of experts from many fields to find an alternative for stopping malaria deaths. The combined forces of the foundation team used cross-fertilization to develop a number of radical solutions to the problem. In one of the brainstorming sessions of this group a rocket scientist who works with guided missiles suggested using lasers to shoot down mosquitoes. He hired optical engineers who experimented with blue lasers from ordinary DVD players and programmers created software to guide the lasers. The inventor then put it all together with parts acquired from eBay. "Harmless to humans and wildlife, the laser is so finely calibrated that it can spot a mosquito by its wing vibrations and bring it down with a tiny burst of light," Covey writes. "Perimeter fences equipped with such lasers are capable of defending entire villages from malaria."

An idea generation technique related to cross-fertilization is discussed in Joel Barker's video *Innovation on the Verge*.[9] This technique, the Verge, can be thought of as a combination of cross-fertilization and "Combine" from SCAMPER (discussed earlier in this chapter). The Verge is where different concepts meet. When these differences come together, they act as triggers to generate a new idea. Barker points out that Innovations on the Verge are hiding all around us and it's our job to find the new combinations. Look to fuse widely different concepts. The examples Barker uses include a brown grocery bag meeting colorful wrapping paper at the verge to generate the idea of a gift bag (i.e., a colored shopping bag). The Bic lighter and a computer meet to generate the plug-in fuel cell to power the computer when the battery runs out. These examples also point out that, in almost all cases, *adaptation* is much quicker and more cost-effective than *invention.*

ANALOGY

Generating ideas by **analogy** is an approach that works quite well for many individuals. With this strategy, we look for analogous situations and problems in both related and unrelated areas. To use this technique effectively, of course, it is important that you read and learn about things outside your area of expertise.

Consider the ZigTech athletic shoes developed by Reebok (see www.the shoegame.com/articles/bill-mcinnis-creator-of-zigtech-interview.html). The concept for the shoe was inspired by a children's toy: the Slinky. The Slinky is a toy that has fascinated children for more than 65 years. When you play with a Slinky, you transfer energy back and forth between rings of the spring. So when Bill McInnis, the inventor of the ZigTech line of shoes, was looking to design energy-efficient athletic shoes, the Slinky came to mind. In Bill McInnis's own words, "If you picture a Slinky stretched out on a table and then look at the side profile of ZigTech you can see the resemblance right away. Think of a mechanical spring, which gives you energy return, then take that same spring idea, but build it out of something soft, like foam, which gives your cushioning. That's how we get energy return and cushioning out of the same platform." The analogy approach successfully provided the inspiration for a new line of athletic shoes. Interestingly, Bill McInnis's job before joining Reebok was working on the space shuttle program. Our next analogy comes from the space program.

In the 1960s, scientists realized they had a problem when they recognized that there was no material available that would survive the high temperatures generated on a space capsule's surface during reentry into the Earth's atmosphere. Consequently, a government directive was issued: "Find a material able to withstand the temperatures encountered on reentry." By the early 1970s, no one had produced a suitable material that satisfied the directive, yet we had sent astronauts to the Moon and back. How had this achievement been possible?

The real problem was "How can we protect the astronauts upon reentry?"— not "Find a material that can withstand such high temperatures." Once the real problem was determined, a solution soon followed. One of the scientists working on the project asked a related question: How do meteors eventually reach the Earth's surface without disintegrating completely? Upon investigation of this problem, he found that although the surface of the meteor vaporized while passing through the atmosphere, the inside of the meteor was not damaged. This analogy led to the idea of using ablative materials on the outside of the capsule that would vaporize when exposed to the high temperatures encountered during reentry. Consequently, the heat generated by friction with the Earth's atmosphere during reentry would be dissipated by the heat of vaporization of a material that coated the outside of the space capsule. By sacrificing this material, the temperatures of the capsule's underlying structural material remained at a tolerable level to protect the astronauts. Once the real problem was uncovered, the scientists solved the problem by using analogies and transferring ideas from one situation to another.

To solve problems by analogy, you should follow four steps.[10]

Solving Problems by Analogy

1. State the problem.
2. Generate analogies (this problem is like …).
3. Solve the analogy.
4. Transfer the solution to the problem.

When generating analogies, apply the same rules you did in brainstorming. For example, in the case of the stale cereal (from Chapter 5), we could say, "Keeping the cereal fresh could be thought of as being similar to preserving raw fish in the tropics without a refrigerator and without cooking." How could we preserve fish in those circumstances? We could add lemon or lime juice to make ceviche (pickled fish). So what could we add to the cereal to keep it fresh? We could add a preservative to cereal for an effect similar to adding lemon juice to the fish. The following example details another situation where an industrial problem was solved by analogy.

Leaking Flow Meter[11]

A flow meter was installed in a chemical plant to measure the flow rate of a corrosive fluid. A few months after its intstallation, the corrosive fluid ate through the meter and began to leak. An extensive time-consuming search was carried out to find a material for the meter that would not corrode, but none was found.

Step 1. State the Problem. (What is the situation?)

The corrosive fluid eats through the flow meter, causing leaks.

> ### Step 2. Generate Analogies. (What else is like this situation?)
>
> Generate as many possibilities as you can, then choose one to work with: Corrosive fluid eating through its meter is like …
>
>> Erosion of a river bank
>> Deterioration of drill bits in the mining industry
>> Paint being worn off an outdoor wooden deck
>
> ### Step 3. Solve the Analogy.
>
> Diamond miners have to replace their drill bits regularly because there is no material harder than the diamonds they are mining.
>
> ### Step 4. Transfer the Solution to the Problem.
>
> The solution used was to develop a schedule to regularly replace the flow meter before the corrosive fluid ate through it to cause a leak; this idea was achieved from the example of the diamond miners replacing their drill bits regularly before they became too dull.

INCUBATING IDEAS

The incubation period is very important in problem solving. Working on a solution to a problem and being forced to meet a deadline often causes you to pick the first solution that comes to mind and then "run with it," instead of stopping to think about alternative solutions. Many times it is advantageous to take a break when working on a problem to let your ideas incubate while your subconscious works on it. Of course, you shouldn't turn the responsibility over to your subconscious completely by saying, "Well, my subconscious hasn't solved the problem yet."

Once the generation of ideas has halted (or you collapse from the effort), an incubation period may be in order. Little is truly understood about mental incubation, but the basic process involves stopping active work on the problem and letting your subconscious continue the work "behind the scenes." Everyone has, at one time or another, been told to "sleep on a problem," in hopes that the solution will be apparent in the morning. This incubation—that is, subconscious work—has been described as a mental scanning of the billions of neurons in the brain in search of a novel or innovative connection to lead to a possible solution.[12]

When members of the National Academy of Engineering were asked, "What do you do when you get stuck on a problem?" some of their responses were as follows:

- "When I can afford the liberty of doing so, I will put the problem down and do something else for a while. My mind keeps working on the problem, and often I will think of something while not trying to."
- "Communicate with other people. Read articles. Try new techniques after a period of digestion. Follow a lead if it looks promising. Keep pursuing."

- "Ask questions about all the circumstances. Ask Socratic questions of yourself. Go home and think. Go to your arsenal of past experiences. Identify factors related to the problem. Read, write, and exchange ideas."
- "I write down everything that I must know to have a solution and everything that I know about the problem so far. Then I usually let it sit overnight, and think about it from time to time. While it is sitting, I often review the recent literature on similar problems and get an idea on how to proceed."

The common thread that runs through these responses is the notion of an incubation period. If the solution to the problem is not an emergency, incubation is a useful (in)activity to consider.

TRIZ

TRIZ, which is pronounced "trees" and stands for "Teoriya Resheniya Izobreatatelskikh Zadatch" and is Russian for "Theory of Inventive Problem Solving (TIPS)," is the brainchild of Russian engineer and scientist Genrich Altshuller.[13–17] Altshuller studied tens of thousands of patents, looking for similarities and innovations in the patents. The TRIZ process recognizes that technical systems evolve so as to increase ideality by overcoming contradictions mostly with minimal introduction of resources. Thus, for creative problem solving, TRIZ provides an algorithmic pathway—that is, it seeks to understand the problem as a system, to imagine the ideal solution first, and to resolve contradictions.

You can think of TRIZ as another way of lateral thinking. It is based on two basic principles, as described by many TRIZ consultants, see Domb:[14]

- Someone, someplace, has already solved your problem or one similar to it. Creativity means finding that solution and adapting it to the current problem.
- Don't accept contradictions. Resolve them.

The first principle essentially tells us, "Don't reinvent the proverbial wheel." The second principle focuses on contradictions. Contradictions include situations in which one object or system has contradictory or opposing requirements, or in which when one feature of the system improves, another worsens. Everyday examples abound:

- Wireless Internet access should be readily available to students on campus (good), but increased access makes it difficult to protect personal information (bad).
- MP3 players should have vast storage capabilities for audio and video (good), but increased storage may increase size and power consumption (bad).

- A product gets stronger (good), but its weight increases (bad). For example, bulletproof vests should be strong (good), but can be bulky and uncomfortable to wear (bad).
- When hard disk space for file storage is increased (good), it becomes more difficult to locate the correct file because the disk is so large (bad).
- Automobile air bags deploy quickly to protect the passenger (good), but the more rapidly they deploy, the more likely they are to injure or kill small or out-of-position people (bad).
- Cell phone networks should have excellent coverage so users have strong signals (good), but cell phone towers are not very nice to look at (bad).
- The email spam filter should be efficient enough to remove all my junk emails (good), but then it is more likely to screen some emails that I actually want to receive (bad).

The first step in the TRIZ process, as described above, is to determine who else has solved a problem like this before. When performing this step, it is helpful to remove the technical jargon from the problem definition and describe it as if you were explaining it to a ten-year-old. This activity may help you identify a similar problem/solution in a disparate field that will aid you in your quest for a solution to the problem at hand. The second step is to identify the ideal solution or, in TRIZ terminology, the *ideal final result* (IFR). The IFR is the solution that you would envision if you had no constraints. The ideal solution is free and magically performs its results perfectly with no negative side effects. The key is to articulate the ideal *results* that you want, without regard to how you arrive at this them at this point— that is, to specify the "What" without regard to the "How." We then strive to achieve a solution as close to the IFR as possible. The third step is to identify the resources that are currently available to solve the problem. For a more thorough discussion of these initial steps in the TRIZ process the reader is referred to Hipple.[17] Many problems can be solved using only these first three steps.

The fourth step of the TRIZ process, and perhaps one of its most interesting features, is to state the problem and the contradictions you are trying to resolve. To resolve the contradictions, we identify which features would improve the solution (good) and which features would worsen the solution (bad). Altshuller assembled a list of 39 features that were common to the thousands of patents he analyzed. A complete description of each of these features can be found on the TRIZ Web site (www.triz-journal.com/archives/1998/11/d/default.asp). A more recent, expanded version of these features (containing 48 in all) is available in Hipple.[17] For our example purposes here, the original 39 will be sufficient to illustrate the process.

---

39 TRIZ Features[15]

1. Weight of moving object	21. Power
2. Weight of stationary object	22. Loss of energy
3. Length of moving object	23. Loss of substance
4. Length of stationary object	24. Loss of information
5. Area of moving object	25. Loss of time
6. Area of stationary object	26. Quantity of substance
7. Volume of moving object	27. Reliability
8. Volume of stationary object	28. Measurement accuracy
9. Speed of object	29. Manufacturing precision
10. Force (intensity)	30. Object-affected harmful
11. Stress or pressure	31. Object-generated harmful
12. Shape	32. Ease of manufacture
13. Stability of the object	33. Ease of operation
14. Strength	34. Ease of repair
15. Durability of moving object	35. Adaptability or versatility
16. Durability of nonmoving object	36. Device complexity
17. Temperature	37. Difficulty of detecting
18. Illumination intensity	38. Extent of automation
19. Use of energy by moving object	39. Productivity
20. Use of energy by stationary object	

This list is used to formulate a 39 × 39 contradiction matrix by listing each of the features along the side and the top of the matrix. The features along the top are those that make the product or situation worse (bad); those along the side improve the product or situation (good).

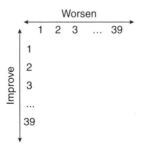

In TRIZ, first we identify the contradictions. Next we use the 39 × 39 contradiction matrix to suggest which of Altshuller's 40 principles (discussed later in this section) might be useful to explore in solving the problem and resolving the contradiction. We formulate the contradiction by specifying which features are improving and which features are worsening. For example, if we were trying to develop a new structural material to be used in airplanes, we could say that we want more strength (improving feature) without more weight (worsening feature). In this case, strength (feature 14) is improving, but the weight of a moving object (feature 1) is worsening. We then locate the improving feature in the first column (vertical) of the matrix and the worsening feature in the first row (horizontal). In our example, the intersection is found in row 14, column 1 of the following abbreviated (14 × 6) contradiction matrix. An interactive form of the full 39 × 39 TRIZ contradiction matrix is available at www.triz40.com/. An expanded version of the contradiction matrix using 48 features is available in Hipple.[17] The expanded matrix contains no blank fields, and therefore should be more helpful in addressing a wider variety of problem contradictions.

Consider a case where (1) weight of the moving objects is a worsening feature and (14) strength is an improving feature. The four numbers in this intersection box are 1, 8, 15, and 40. These four numbers refer to four of the 40 Altshuller principles that may help us solve the problem, which in this case are the following:

1 (segmentation)
8 (anti-weight)
15 (dynamics)
40 (composite material films)

We now brainstorm solutions using the Altshuller principles to resolve our contradiction and solve our current problem (e.g., developing a better airplane material). Altshuller developed these 40 principles in an effort to summarize solution techniques that have already been identified for other problems (i.e., someone, someplace, has already solved a similar problem). Explanations of each of the 40 principles are available on the TRIZ40 Web site.[15]

TRIZ Contradiction Matrix[15]

		Worsening Feature					
		1. Weight of moving object	2. Weight of stationary object	3. Length of moving object	4. Length of stationary object	5. Area of moving object	6. Area of stationary object
Improving Feature	1. Weight of moving object	*	—	15, 8 29, 34	—	29, 17 38, 34	—
	2. Weight of stationary object	—	*	—	10, 1 29, 35	—	35, 30 13, 2
	3. Length of moving object	8, 15 29, 34	—	*	—	15, 17 4	—
	4. Length of stationary object	—	35, 28 40, 29	—	*	—	17, 7 10, 40
	5. Area of moving object	2, 17 29, 4	—	14, 15 18, 4	—	*	—
	6. Area of stationary object	—	30, 2 14, 18	—	26, 7 9, 39	—	*

Continues

	Worsening Feature					
	1. Weight of moving object	2. Weight of stationary object	3. Length of moving object	4. Length of stationary object	5. Area of moving object	6. Area of stationary object
7. Volume of moving object	2, 26 29, 40	—	1, 7 4, 35	—	1, 7 4, 17	—
8. Volume of stationary object	—	35, 10 19, 14	19, 14	35, 8 2, 14	—	—
9. Speed of object	2, 28 13, 38	—	13, 14 8	—	29, 30 34	—
10. Force (Intensity)	8, 1 37, 18	18, 13 1, 28	17, 19 9, 36	28, 10	19, 10 15	1, 18 36, 37
11. Stress or pressure	10, 36 37, 40	13, 29 10, 18	35, 10 36	35, 1 14, 16	10, 15 36, 28	10, 15 36, 37
12. Shape	8, 10 29, 40	15, 10 26, 3	29, 34 5, 4	13, 14 10, 7	5, 34 4, 10	—
13. Stability of the object	21, 35 2, 39	26, 39 1, 40	13, 15 1, 28	37	2, 11 13	39
14. Strength	1, 8 40, 15	40, 26 27, 1	1, 15 8, 35	15, 14 28, 26	3, 34 40, 29	9, 40 28

The left side of the table is labeled **Improving Feature**.

Altshuller's 40 Principles of TRIZ[15]

1. Segmentation
2. Taking out
3. Local quality
4. Asymmetry
5. Merging
6. Universality
7. "Nested doll"
8. Anti-weight
9. Preliminary anti-action
10. Preliminary action
11. Beforehand cushioning
12. Equipotentiality
13. The other way around
14. Spheroidality
15. Dynamics
16. Partial or excessive actions
17. Another dimension
18. Mechanical vibration
19. Periodic action
20. Continuity of useful action
21. Skipping
22. "Blessing in disguise"
23. Feedback
24. "Intermediary"
25. Self-service
26. Copying
27. Cheap short-living
28. Mechanics substitution
29. Pneumatics and hydraulics
30. Flexible shells and thin films
31. Porous materials
32. Color changes
33. Homogeneity
34. Discarding and recovering
35. Parameter changes
36. Phase transitions
37. Thermal expansion
38. Strong oxidants
39. Inert atmosphere
40. Composite material films

Now let's reconstruct how TRIZ was applied to solve the following problem that Boeing Aircraft Company faced several years ago.[17]

Increasing the Size of the Boeing 737

After the initial success of the Boeing 737-100 and 737-200 series it was decided to increase the capacity from approximately 100 passengers to 140 passengers by making the plane longer by approximately 10 feet but by keeping other dimensions essentially the same. This larger-capacity aircraft would require larger physical-size engines to accommodate greater air intake as well as a larger cowl that houses the fuel injection lines surrounding the intake. Initial designs showed that when the larger engines were to be placed on the wings, the clearance from the bottom of the engine to the ground was too small to meet the safety requirements required for the possibility of an occasional bumpy landing. Neither the wing nor engine height could be raised without substantial redesign and testing. Use the steps in TRIZ procedure to suggest a solution for this Boeing 737 problem.

Source: Photo copyright Ivan Cholakov/Shutterstock.

Boeing 737-200

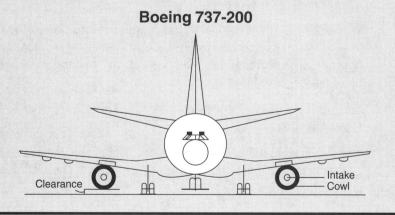

TRIZ Contradiction Matrix

	Worsening Feature				
	1. Weight of moving object	2. Weight of stationary object	3. Length of moving object	4. Length of stationary object	5. Area of moving object
1. Weight of moving object	*	—	15, 8 29, 34	—	29, 17 38, 34
2. Weight of stationary object	—	*	—	10, 1 29, 35	—
3. Length of moving object	8, 15 29, 34	—	*	—	15, 17 4
4. Length of stationary object	—	35, 28 40, 29	—	*	—
5. Area of moving object	2, 17 29, 4	—	14, 15 18, 4	—	*
6. Area of stationary object	—	30, 2 14, 18	—	26, 7 9, 39	—
7. Volume of moving object	2, 26 29, 40	—	1, 7 4, 35	—	1, 7 4, 17

(Improving Feature is labeled vertically along the left side of the rows.)

Example Application of TRIZ

Problem Statement: Engines should be larger and clearance remain unchanged.

Step 1.　Identify the Contradiction(s)

We need the airplane engine air intake and fuel injection casing to be larger to process and burn more fuel to fly the larger aircraft. The larger engines decrease the clearance (length), making it unacceptable to landing standards.

Step 2.　Look at the List of Features and Identify Those Important to Your Contradiction

The airplane engine is a moving object. The air intake area and the cowl with the fuel lines make up the engine *volume*. The TRIZ feature is *the volume of moving object*.

The larger engine will decrease the *length* of the clearance. The TRIZ feature is the *length of the moving object*.

Step 3. *Identify Which Are Improving Features and Which Are Worsening Features*

The improving feature is number 7 "Volume of the moving object" and the worsening feature is number 3 "length (clearance) of the moving object" in the contradiction matrix.

Step 4. *Refer to the TRIZ Contradiction Matrix to Learn Which of Altshuller's Principles May Be Useful for the Problem*

We can now go to the TRIZ Web site (www.triz40.com/) to find that the intersection of these two contradictions, 7 and 3, give the following Altshuller principles:

1. Segmentation

4. Asymmetry

7. "Nested doll"

35. Parameter changes

Step 5. *Brainstorm how we could use each of these four principles to solve our problem.*

1. Segmentation

Here is the explanation of this principle from the TRIZ Web site:

- Divide an object into independent parts.
- Make an object easy to disassemble (replace worn or damaged parts).
- Increase the degree of fragmentation or segmentation.

We have the engine air intake area and the area of the casing or cowl surrounding the intake. The intake area must be circular because of the spinning blades inside the engine.

4. Asymmetry

Here is the explanation of this principle from the TRIZ Web site:

- Change the shape of an object from symmetrical to asymmetrical.
- If an object is asymmetrical, increase its degree of asymmetry.

Does the cowl surrounding the air intake have to be symmetric, that is, circular? No, it does not!

Continues

7. "Nested doll"

Here is the explanation of this principle from the TRIZ Web site:

- Nesting by placing one object inside another; place each object, in turn, inside the other.
- Make one part pass through a cavity in the other.

Could the symmetrical spinning blade area be "nested" inside an asymmetrical casing? Could the bottom of the cowl be flattened and thus leave a greater clearance?

35. Parameter changes

This principle does not appear to be readily applicable to this problem. This is not necessarily unusual, because Altshuller's principles are merely general suggestions intended to help us focus our thinking in areas that have proven fruitful in previous problems.

The solution adopted by the Boeing engineers was to nest the symmetrical intake inside an asymmetrical casing as shown in the following schematic.

Source: Photo copyright Arena Photo UK/Shutterstock.

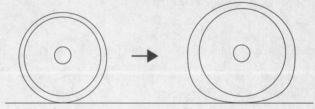

Source: Photo copyright Kvini/Shutterstock.

Thus, by identifying problem contradictions, we can use the TRIZ process to help us reach a solution. In this case, we were able to generate two additional ideas using the TRIZ method.

It is interesting to note that we could have arrived at the same solution based on asymmetry had we selected "Area of a moving object" instead of "Volume of a moving object" as the improving feature. This is to be expected since volume and area are directly related.

USE THE DEFECT TO SOLVE THE PROBLEM

Another tenet of TRIZ is the use of defects to solve the problem. A defect is a flaw in the product or process that needs to be eliminated. TRIZ asks the question "How can one use this defect in a positive way to solve the problem?" The following example outlines a soap manufacturer's ability to do just that.

Empty Soapboxes

A manufacturer of soap had been receiving complaints from its customers that some of its soapboxes were empty (i.e., the defect or negative resource—lack of weight). The company, wanting to be sure that all boxes shipped contained a bar of soap, devised a high-tech solution where each box would be X-rayed, and empty boxes would then be removed by a robotic arm as shown in the figure.

Continues

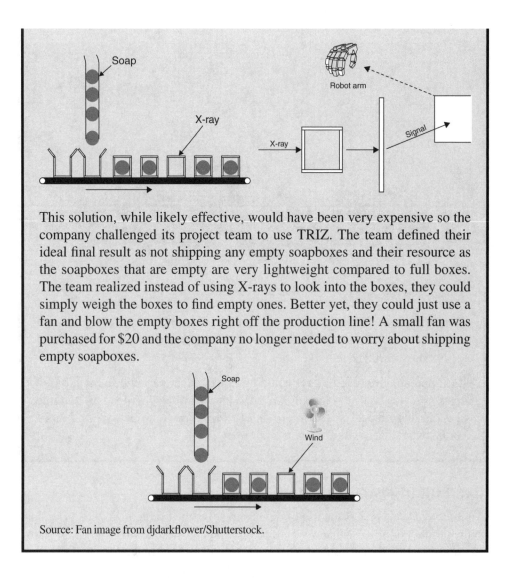

This solution, while likely effective, would have been very expensive so the company challenged its project team to use TRIZ. The team defined their ideal final result as not shipping any empty soapboxes and their resource as the soapboxes that are empty are very lightweight compared to full boxes. The team realized instead of using X-rays to look into the boxes, they could simply weigh the boxes to find empty ones. Better yet, they could just use a fan and blow the empty boxes right off the production line! A small fan was purchased for $20 and the company no longer needed to worry about shipping empty soapboxes.

Source: Fan image from djdarkflower/Shutterstock.

A more advanced form of TRIZ is called Algorithm of Inventive Problem Solving (ARIZ). ARIZ consists of a number of procedures to solve more complex problems, where the fundamental tools of TRIZ alone are not adequate. Various TRIZ software tools make use of this algorithm. If the tools of TRIZ are used in an effective manner, the major challenges of today will be resolved more rapidly to produce the success stories of tomorrow.[18]

The authors would like to thank Glenn Mazur and Jack Hipple for introducing them to this topic and to Jack Hipple for providing the examples.

SUMMARY

This chapter has presented techniques to help you generate creative solutions. Although it can be advantageous to take a break when working on a problem to let your ideas incubate, don't turn the entire responsibility over to your subconscious. Instead, use the following techniques to help spur your creativity:

- Use free association brainstorming to generate solutions to the problem.
  - Use vertical thinking (SCAMPER) to build on previous ideas and generate new ideas. If you need even more ideas, use the 77 design heuristics.
  - Use lateral thinking—random stimulation and other people's views—to generate new ideas when you are stuck in a rut.
- Use a *fishbone* diagram to help organize the ideas and solutions you generate.
- Use *futuring* to remove all technical blocks to envision a solution in the future.
- Use *cross-fertilization* to bring ideas, phenomena, and knowledge from other disciplines to bear on your problem.
- Use *analogies* to find solutions to similar problems in other disciplines.
- Use *TRIZ* to resolve contradictions or use the defect to solve the problem.
- Incubate ideas for a while after generating them; then revisit ideas and see if you can improve them.

WEB-SITE MATERIAL (WWW.UMICH.EDU/~SCPS)

- **Learning Resources**
  Summary Notes
- **Interactive Computer Modules**

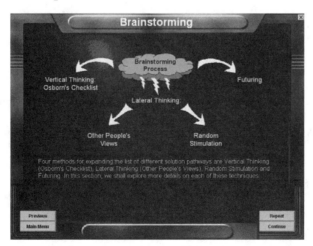

Brainstorming

- **Professional Reference Shelf**
  1. Overall Scheme of the Brainstorming Process
  2. Vertical and Lateral Thinking
     - Basketball Exercises – The Case of the Putrid Pond – Space Capsule

  3. Futuring
     - Useful Products from Cheese Waste

  4. Analogy and Cross-Fertilization
     - Shockblocker Shoes
     - A Cold Winter's Day
     - Dinner at Antoine's

  5. TRIZ
     - A New Structural Material for Bulletproof Garments

REFERENCES

1. SCAMPER was defined by Robert Eberle and is a slight modification of Osborn's checklist. See www.mindtools.com/pages/article/newCT_02.htm.
2. Daly, S., Yilmaz, S., Christian, J., Seifert, C. M., and Gonzalez, R., "Design Heuristics in Engineering Concept Generation," *Journal of Engineering Education*, 101 (4), pp. 601–629, 2012.
3. Yilmaz, S., and Seifert, C. M., "Creativity Through Design Heuristics: A Case Study of Expert Product Design," *Design Studies*, 32 (4), pp. 384–415, 2011.
4. Daly, S., Christian, J., Yilmaz, S., Seifert, C., and Gonzalez, R., "Assessing Design Heuristics in Idea Generation within an Introductory Engineering Design Course," *International Journal of Engineering Education*, 28 (2), pp. 463–473, 2012.

5. Yilmaz, S., Christian, J. L., Daly, S. R., Seifert, C. M., and Gonzalez, R., "Collaborative Idea Generation Using Design Heuristics," in 18th International Conference on Engineering Design, ICED '11, Copenhagen, Denmark, 2011.

6. De Bono, E., *Lateral Thinking: Creativity Step by Step*, Harper & Row, New York, 1970.

7. De Bono, E., *Serious Creativity: Using the Power of Lateral Thinking to Create New Ideas*, Harper Business, a division of HarperCollins Publishers, New York, 1993.

8. Covey, Steven R., *The 3rd Alternative: Solving Life's Most Difficult Problems*, Breck England, Simon & Schuster, October 4, 2011.

9. Barker, J. A., *Discovering the Future: The Business of Paradigms*, ILI Press, St. Paul, MN, 1985; *The New Business of Paradigms, 21st Century Editions*, distributed by Star Thrower (800-242-3220); *Paradigm Pioneers and the Power of Vision, 21st Century Editions*, distributed by Star Thrower (800-242-3220). Also see http://abcnews.go.com/Technology/story?id=1518077#.TzmFJhyfeME and www.npr.org/2010/11/08/131163403/its-time-the-wristwatch-makes-a-comeback.

10. Van Gundy, A. B., *Techniques of Structured Problem Solving*, 2nd ed., Van Nostrand Reinhold, New York, 1988.

11. McNally, Dr. R. G., Dow Chemical Company, Midland, MI 48667.

12. Reid, R. C., "Creativity?" *Chemtech*, 17, p. 14, January 1987.

13. Altshuller, G. S., and R. V. Shapiro, "About a Technology of Creativity," *Questions of Psychology*, pp. 37–49, 1956.

14. Domb, Ellen, "Think TRIZ for Creative Problem Solving," *Quality Digest*, 2003, www.qualitydigest.com/aug05/articles/03_article.shtml.

15. Altshuller, G., *40 Principles: TRIZ Keys to Innovation*, extended edition, Worchester, MA, Technical Innovation Center, Inc., April 2005. ISBN: 09640740-5-2.

16. Mazur, Glenn H., Executive Director of QFD Institute, Ann Arbor, MI, www.mazur.net.

17. Hipple, Jack, *The Ideal Result—What It Is and How to Achieve It*, Springer Science+Business Media, New York 2012.

18. www.triz-journal.com/archives/1998/11/d/default.asp.

19. Hipple, Jack, TRIZ and Engineering Training Services LLC (www.innovation-triz.com).

EXERCISES

7.1. Make a list of the following:

A. The five most important things you learned from this chapter

B. The five most interesting things you learned from this chapter that you could use to make a conversation more interesting

7.2. Load and run the Interactive Computer Module for Brainstorming from the SCPS Web site. Record your performance number:_____ _____.

7.3. Keep a journal of all the good ideas you generate.

7.4. A. Make a list of the worst business ideas you can think of (e.g., a maternity shop in a retirement village, a solar-powered night-light, reversible diapers).

B. Take the list you generated in part (a) and turn it around to make them viable concepts for entrepreneurial ventures (e.g., reversible diapers—blue on one side and pink on the other).

7.5. Apply the following four steps for generating solutions to a problem you have by analogy.

    A. State the problem: _____

    B. Create analogies: This situation is like _____

    C. Solve the analogy: _____

    D. Transfer the solution: _____

7.6. Watch only the first half of a movie or TV show with a friend. Have each of you "create" your own ending. Watch the rest and discuss the results. Whose ending was better? Why?

7.7. Write a paragraph discussing how you can improve your ability to generate ideas. Compare and contrast the various idea generation techniques. Is there a common thread that runs through all of these techniques? Identify situations in which each technique might best apply.

7.8. Suggest 50 ways to increase spectator participation at (a) professional basketball games. [Examples: Have a drawing at each game in which the people in the randomly selected seats get to play for two minutes. Give the fans one arrow each to shoot at the basketball in midair to try to block the shot.] Now suggest 25 ways for spectator participation in (b) football, (c) baseball, and (d) hockey.

7.9. You are a passenger in a car that lacks a speedometer. Describe 25 ways to determine the speed of the car.

7.10. An epidemic on a chicken farm created 1,000 tons of dead chickens. The local landfill would not accept the dead chickens for disposal. It is also against the law to bury the chickens. The local authorities are insisting that the matter be dealt with immediately. Suggest ways to solve the farmer's problems.
Problem adapted from *Chemtech*, 22, 3, p. 192, 1992.

7.11. A reforestation effort in Canada is running into trouble in a particular region. In one nursery alone, 10 million seedlings were eaten by voles. The voles even consumed the varieties chosen for the unpalatable phenol/condensed tannin secondary metabolite they contain. The voles overcame this unpalatability by cutting the branches, stripping the bark, and then leaving the branches for a few days before eating them. This process caused the unpleasant components to decline to acceptable levels. Suggest 15 ways to solve the reforestation problem in this nursery.
Problem adapted from *Chemtech*, 21, p. 324, 1991.

7.12. Kite flying is a growing hobby around the world. (Kites are very entertaining—it is not unusual to find kites that fly at altitudes of more than 2,000 feet.) Suggest 25 ways that kites can be used for purposes other than entertainment.

7.13. The use of a steam cycle is a popular means of generating electricity for industrial and domestic use. Unfortunately, the current theoretical maximum efficiency of a steam plant is approximately 40% and the effects on the environment of emissions from these plants are of growing concern. How do you envision energy being produced and consumed in the future?

7.14. Choose two people from different professions (e.g., repair person, florist, dentist, accountant, police officer, hockey coach, car designer, custodian, bellhop, cruise ship activity director, Cub Scout leader) and make lists similar to the ones below suggesting what these individuals could learn from each other that would enrich each other's lives.

Problem courtesy of Matt Latham and Susan Stagg Williams.

Pastor Gives to a Maitre d'

A. Ideas to rapidly assess people's needs

B. Suggestions on how not to take every problem she hears personally (thick-skinned)

C. The importance of a well-groomed physical appearance

D. Suggestions on how far you can push people (in terms of views and ideals)

E. Ideas on offering suggestions and advice

F. Ideas on how to be more self-reliant (scheduling)

Maitre d' Gives to a Pastor

A. Knowledge to calm upset individuals or perform crowd control

B. Understanding and dealing with people; approachability

C. Memory techniques to remember frequent customers

D. An appreciation of having a boss and someone watching what you do

E. Ideas on how to learn to be happy with your job and yourself

7.15. Read the article by J. M. Prausnitz (Professor at the University of California–Berkeley) entitled "Toward Encouraging Creativity in Students" in *Chemical Engineering Education* (Winter 1985, p. 22), http://cee.che.ufl.edu.

A. Discuss the ideas presented on problem recognition.

B. How are creativity and synthesis defined?

C. Discuss two examples used to illustrate "the tying together of two separate ideas."

D. What does the article suggest about creativity and the cross-fertilization of ideas from one area to a completely different area?

E. What is the single most important point the author is making about problem solving?

7.16. Make a list of several ways you can improve your creative abilities. Describe how you would implement some techniques from the table on pages 133–135.

7.17. Carry out a futuring exercise to visualize the following items:

    A. A telephone call in the year 2020

    B. Eating a meal with your family in the year 2050

    C. A homework assignment in the year 2025

    D. A homework assignment in the year 2125

7.18. Describe the most creative television advertisement you have seen (Super Bowl commercials are a good source), and explain why it was so creative.

7.19. Use the TRIZ method to generate possible solutions for these problems:

    A. Wireless Internet access should be readily available to students on campus, yet secure enough to protect personal information.

    B. The email spam filter should be efficient enough to remove all of your junk emails (good), but then it is more likely to screen some emails that you actually want (bad).

    C. Increase the level of services provided by the U.S. Postal Service without increasing the cost.

    D. Figure out a way for me to lose 10 pounds, without having to be hungry all the time.

7.20. Carry out a TRIZ analysis for the following problems:

    A. Automobiles should be strong and sturdy enough to provide adequate crash protection, yet light enough to allow for good gas mileage.

    B. Bulletproof vests should be strong (good), yet not too bulky and uncomfortable to wear (bad).

    C. Increased hard disk space allows for more file storage (good), but creates difficulty in locating the correct file because the disk is so large (bad).

    D. Automobile air bags deploy quickly to protect the passenger (good), but the more rapidly they deploy, the more likely they are to injure or kill small or out-of-position people (bad).

    E. Cell phone networks should have excellent coverage so that users have strong signals (good), but cell phone towers are not very nice to look at (bad).

7.21. TRIZ: Electricity should be produced in large quantities for efficiency, but not transported far due to losses in transmission. Examine the situation by using the TRIZ contradiction matrix to come up with possible solutions.

7.22. TRIZ: Smartphones should have large screens for use but should be small and light so they are easy to carry around. Examine the situation by using the TRIZ contradiction matrix to come up with possible solutions.

7.23. A pill manufacturer is having issues with some pills being in the shape of spheres instead of ovals. Brainstorm ways to use the defect to solve the problem of shipping spherical pills.

7.24. A village called Dull in Perthshire, England, is about to become a sister city with a town called Boring in Oregon. Brainstorm the title of the yearly festival to celebrate this union and what events would you hold at the festival. Use free association, then SCAMPER, followed by random simulation to generate your ideas. Pick the best two titles and the five best events.

Source: *London Metro* (Times), April 25, 2012, page 7.

7.25. "Incubating Ideas": Take your answer to any of the exercises in this chapter and try to improve on it the next day. Come up with five improvements you can make to your original solution.

7.26. "Gas Well Liquid Loading": In hydraulic fracturing (fracking), water is pumped into shale rock formations to break up rock and allow natural gas trapped in the rock to flow out of the well.

During fracking, the more water that is pumped into the well, the greater the amount of rock that is fractured, allowing access to more gas. However, as more water is pumped, it can block the flow of natural gas in the well and reservoir from getting to the surface.

Using www.triz40.com, identify the TRIZ contradiction, find which of Altshuller's principles may be used for the problem, and propose solutions.

Based on real field engineering.

FURTHER READING

Adams, James L. *Conceptual Blockbusting: A Guide to Better Ideas*, 3rd ed. Addison-Wesley, Stanford, CA, 1986.

Mihalko, Michael. *Cracking Creativity: The Secrets of Creative Genius*. Ten Speed Press, Berkeley, CA, 2001.

Mihalko, Michael. *Thinkertoys*, 2nd ed. Ten Speed Press, Berkeley, CA, 2006.

Von Oech, Roger. *A Whack on the Side of the Head: How You Can Be More Creative*, revised edition. Warner Books, New York, 1990.

www.aitriz.org, The Altschuller Institute for TRIZ Studies.

www.asme.org/products/courses/triz--the-theory-of-inventive-problem-solving, TRIZ Short Course offered through ASME.

8   DECIDING THE COURSE OF ACTION

Once the real problems are defined and solutions are generated, it is time to apply the third step in our heuristic: **decide**. Specifically, we must do the following:

- Decide which problem to address first and how to address it.
- Decide the best solution from our possible solution alternatives.
- Decide how to avoid additional problems as we implement our chosen solution.

An organized process for making these essential decisions is the Kepner–Tregoe (K.T.) approach, which is described in *The New Rational Manager* and shown in the following diagram.[1,2]

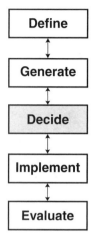

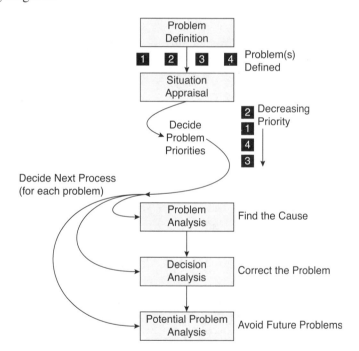

Components of the Kepner–Tregoe Approach

After we *define the problem(s)* we move through the K.T. approach by starting with a K.T. *situation appraisal*, which helps us when we are faced with multiple problems at the same time. During a situation appraisal, we do two things: (1) analyze the problems that we are facing to determine their importance and set the priority (based on several criteria that we will describe) and (2) classify the problem by the type of action that will be required for solution.

K.T. SITUATION APPRAISAL

Deciding the Priority

In many situations, a number of problems arise at the same time. In some cases they are interconnected; in other cases, they are totally unrelated. When faced with multiple problems, a K.T. situation appraisal can prove useful in deciding which problem receives the highest priority.

We first make a list of all of the problems and then try to decide which problem in this group should receive attention first. The priority of each problem will be evaluated using the following criteria: *timing, trend,* and *impact.* The criteria are rated as having either a high degree of concern (H), a moderate degree of concern (M), or a low degree of concern (L).

Evaluation Criteria

1. *Timing:* How urgent is the problem? Is a deadline involved? What will happen if nothing is done for a while? For example, suppose you are the manager of a bakery and one of the five ovens in a bakery is malfunctioning, but the other four ovens could pick up the extra load. It may be possible to wait on this problem and address more urgent problems, so we would give the problem an L rating (low degree of concern). On the other hand, if the other four ovens are operating at maximum capacity and a major order must be filled by the evening, the rating for *timing* would be H (high degree of concern) because the problem must be solved now.

2. *Trend:* What is the problem's potential for growth? In the bakery example, suppose the malfunctioning oven is overheating, getting hotter and hotter, and cannot be turned off. Consequently the *trend* is getting worse, and you

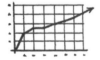

   have a high degree of concern (an H) about a fire starting. You also could have a high degree of concern if you are getting further and further behind on your customers' orders. On the other hand, if the oven is off and you can keep up with the orders with four ovens, the *trend* is a low degree of concern (L).

3. *Impact:* How serious is the problem? What are the effects on the people, the product, the organization, and its policies? In the bakery example, suppose you cannot get the oven repaired in time to fill the order of a major client. If, as a result, you could subsequently lose the client's business, then the *impact*  is a high degree of concern (H). On the other hand, if you can find a way to fill all the orders for the next few days, then the *impact* of one malfunctioning oven is a moderate degree of concern (M).

Deciding the Action to Take

In the second part of a situation appraisal we ask the following questions to determine the action we should take:

- Do we need to learn the cause of the problem? If so, it would lead us to use the *K.T. problem analysis*, or PA. In a problem analysis, the cause of the problem or the fault is unknown and we have to find it. What is it that happened in the *past* that is causing the current trouble? The Kepner–Tregoe approach for finding the root cause was discussed in Chapter 5. Once the problem is known, we need to make a decision of what to do about the problem so we move to a *K.T. decision analysis* (KTDA), where we make a decision on what to do about the problem.

or

- Do we need to make a decision among possible alternatives that face us? For this we would use a KTDA. With a KTDA, the cause of the problem has been found and now we need to decide what to do about it. The decision at the *present* time is how to correct the problem. Once the decision is made, we move to K.T. *potential problem analysis* where we take steps to ensure the success of our decision.

or

- Do we need to analyze the potential problems that might arise from our solution alternatives, and then take preventative actions and make some contingency plans to ensure the success of our solution? To address this, we would use *potential problem analysis* (KTPPA). With a KTPPA, we want to ensure the success of our decision and anticipate and prevent *future* problems from occurring.

We now consider several other examples and solutions to help illustrate the K.T. approach to prioritizing problems.

K.T. Situation Appraisal:
You Know It's a Really Bad Day When …

First let's consider the problem of the man pictured below. Here we see a dog hanging on to the homeowner's leg as he clings to the briefcase containing very important papers while his house burns. His car is not working and it appears that in the distance a tornado is approaching.

We see that we have five problems to address in filling out our situation analysis table.

Problem	Timing (H, M, L)	Trend (H, M, L)	Impact (H, M, L)	Next Process (PA, DA, PPA)
1. Get dog off leg				
2. Repair car				
3. Put out fire				
4. Protect papers in briefcase				
5. Prepare for touchdown of tornado				

Let's discuss each problem and evaluate the criteria for each.

1. It is necessary to get the dog off your leg now (high priority). The trend is getting worse because wounds from the bite are becoming more serious (H) and the impact is that you can do nothing else until the dog is off your leg (H). The process is to decide how to get the dog off your leg (DA).

2. Repairing the car can wait (low priority) and it is not getting worse (L), but if it is not repaired soon it could impact your job by your not being able to visit clients (moderate priority). The problem is to find out what is wrong with the car (PA).

3. Putting out the fire receives high priority in all three categories. The problem is to decide (DA) how to do it: Get the hose or fire extinguisher; call the fire department; and/or make sure everyone is out of the house.

4. If you rush off to handle the other projects in this list, you need to make sure your months of work, which includes signed documents in your briefcase, are protected. The process is one of PPA and of making sure your signed papers (which your clients now wish they had not signed) are in a safe place.

5. While the tornado looks somewhat close in the picture, it may be used to represent a tornado in the area, and thus may only be a tornado warning. So this hazard could merit a KTPPA.

Problem	Timing (H, M, L)	Trend (H, M, L)	Impact (H, M, L)	Next Process (KTPA, KTDA, KTPPA)
1. Get dog off leg	H	H	H	KTDA
2. Repair car	L	L	M	KTPA
3. Put out fire	H	H	H	KTDA
4. Protect papers in briefcase	M	M	H	KTPPA
5. Prepare for touchdown of tornado	M	H	H	KTPPA

We see we have two problems that have high (H) ratings in all categories and we have to decide which we shall work on first. To do this, we do a **pairwise comparison** of these two problems in each category. In the *timing* category the person cannot really move to put out the fire or call the fire department because of the dog on the leg, so that project gets the vote in the timing category. In the trend category the dog's teeth are going deeper and deeper into the leg so the pain is such that the person cannot really think about anything else, so the dog gets the top vote in this category as well. The dog also gets the top vote in the impact category because the person cannot move to do anything else. So the first task is to decide how to get the dog off the person's leg.

Maen CG/Shutterstock

K.T. Situation Appraisal: Flooding in Thailand

It is late summer 2011, and Thailand is experiencing one of the worst floods in its history with water levels reaching 16-year highs. Throughout the year, heavy precipitation occurred in the northern areas of Thailand, leading to flooding in many provinces. The flooding is still expanding south toward the capital city of Bangkok as rain continues to pour down.

The government is faced with limited resources and needs to decide what to prioritize as aid comes in from around the world. Specifically, a new group of emergency helicopters is available and could be tasked with supplying food and water to the affected areas in the North, evacuating the people from the areas that are currently flooding, or transporting sandbags to protect the remaining areas in the South, where new projections show flooding will reach in the not-too-distant future.

Where should the government send the helicopters first? The government narrowed the use of the helicopters to the following three choices: (1) to transport food and water, (2) to evacuate people, and (3) to transport sandbags.

1. Food, water, and other supplies are very important for people to survive and are in short supply in the affected areas. The government is already using planes to send water, food, and supplies to some impacted regions, but the flood has spread to many areas that are only accessible by helicopter. Consequently, the impact of not getting food and water to these areas is high (H). Some of the areas unreachable by plane have been flooded for months now and the food must be brought in *now*, so timing is a high degree of concern (H). More and more areas are running short of supplies as the flooding spreads and remains in regions for extended periods of time. The list of areas that need shipments will only continue to grow so the trend is a high degree of concern (H). The next step is to analyze which areas need what supplies and how soon they need them (KTPA).

2. The flood has moved from the North to the South over a period of months and many people decided to evacuate before it hit their area. However, some stubborn individuals remained and are now in dire need of evacuation as they are stranded on tops of buildings as floodwaters

rise (timing = H). If not rescued soon, these people could drown because floodwaters have reached heights of five meters in some places (trend = H). In the regions where flooding is reaching heights significant enough to strand people, boats are already being used to rescue the people, but the helicopters would be much more efficient as the water is full of debris, making it tough to safely maneuver boats (impact = M). The process to address this issue is a decision analysis of how and where to use both boats and helicopters to save all people.

3. Generally, most floods in Thailand do not reach the South, but this year it is expected to reach Bangkok and as a result the city needs to construct sandbag walls to protect its inhabitants and infrastructure. The flood moves slowly but steadily and is expected to reach Bangkok in the next couple of weeks. It should only take a few days to create significant barriers (timing = M). However, if the sandbags are not in place before the flood arrives, significant damage will be done to the city and lives could be lost (trend = H). Helicopters could transport sandbags to build dams to block the water streams from reaching the cities and urban areas, thereby potentially saving lives and preventing significant damage to buildings. However, because the areas are not yet flooded, large trucks can also be used to do the job (impact = L). The next step is to brainstorm what could go wrong once the sandbags are in place (KTPPA).

Projects	Timing (H, M, L) *How urgent is the problem? What will happen if nothing is done?*	Trend (H, M, L) *What is the problem's potential of growth?*	Impact (H, M, L) *How serious is the problem? What are the effects on the people?*	Next Process (KTPA, KTDA, KTPPA)
1. Transport food and water supplies	H	H	H	KTPA
2. Evacuate flooding areas	H	H	M	KTDA
3. Transport sandbags	M	H	L	KTPPA

After summarizing the results from the previous discussions, it can be seen, in the table above, that the government should first send the new helicopters they have to transport food and water to the affected people in the northern areas.

K.T. PROBLEM ANALYSIS

After completing *the situation analysis* we will either begin working on a problem to solve (i.e., the KTPA), making a decision (i.e., the KTDA), or deciding how to

ensure the success of the alternative we have chosen (i.e., the KTPPA). We already discussed the K.T. problem analysis in Chapter 5. With the KTPA, we looked at the four dimensions of the problem: *what, where, when,* and *extent,* and we asked what IS and IS NOT and what is the *distinction* between the two.

K.T. Problem Analysis

	Is	Is Not	Distinction	Probable Cause
What				
Where				
When				
Extent				

We will also use K.T. problem analysis in the Woods troubleshooting algorithm in Chapter 11.[3]

K.T. DECISION ANALYSIS

In this section, we will discuss *how to choose the best solution* from a number of alternative solutions that have been formulated to solve the problem. The K.T. deci-

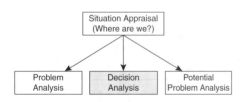

sion analysis (KTDA) is an algorithm for choosing between different alternative solutions to find the one that best fulfills all the objectives. The KTDA is a way of organizing and displaying the data by weighing the desired objectives and evaluating how each of the alternatives fulfills each objective. The following blank format shows the process for using this algorithm.

1. Write a concise **decision statement** about what it is you want to decide and then use the first four steps discussed in Chapter 4 to gather information. Identify what you are trying to accomplish and what resources are available.
2. List all the objectives to be achieved by the decision and divide these objectives into two categories: **musts** and **wants.** The musts are mandatory to

<table>
<tr><td colspan="2">Decision Analysis</td></tr>
<tr><td>Decision Statement: _____</td></tr>
<tr><td>_____</td></tr>
</table>

achieve a successful solution and they have to be measurable and realistic. The wants are desirable but not mandatory. For example, if you are buying a new shirt and only have $20 to spend, a must would be that it is $20 or less. A want might be that it is blue, something you desire, but isn't an absolute.

3. Develop a list of alternatives or options (A, B, C) from which to choose.

<table>
<tr><td colspan="2">Musts</td></tr>
<tr><td>1.</td></tr>
<tr><td>2.</td></tr>
<tr><td>3.</td></tr>
<tr><td colspan="2">Wants</td></tr>
<tr><td>1.</td></tr>
<tr><td>2.</td></tr>
<tr><td>3.</td></tr>
</table>

4. Evaluate each alternative solution against each of the ***musts.*** If the alternative

Alternative Solution:	A	B	C

solution satisfies all the musts, it is a "go" and you should continue to evaluate that alternative; if it does not satisfy any one of the musts, it is a "no go," that is, it should not be considered further, as is the case shown below where Alternative B did not satisfy Must 2. Consequently, there is no need to consider this alternative further and we do not need to evaluate the ***wants*** for Alternative B.

Decision Analysis

Decision Statement: _____

_____

Alternative Solution:		A		B		C	
Musts	1.	Go		Go		Go	
	2.	Go		No Go		Go	
Wants	Weight	Rating	Score	Rating	Score	Rating	Score
1.					No Go		
2.					(crossed out)		
		Total A =		Total B =		Total C =	

5. After learning which alternatives satisfy the musts, proceed to examine the list of the objectives you want to satisfy. The wants are desirable but not mandatory and give you a comparative picture of how the alternatives perform relative to each other. We make a list of all the characteristics that the solution should have and then order them by putting the most important want first and the least important last. Next assign a *weight* (1–10) to each want to give a sense of how important that characteristic (or want) is. Assigning weights is indeed a subjective task. If a want is extremely important, it should be given a high weight of 9 or 10. However, if it is only moderately important, the weight should be a lower weight of 6 or 7.

Decision Analysis

Decision Statement: _____

_____

Alternative Solution:		A		C	
Musts	1.	Go		Go	
	2.	Go		Go	
Wants	Weight	Rating	Score	Rating	Score
1.	8				
2.	7				
		Total A =		Total C =	

6. The next step is to evaluate each alternative against the wants and give it a *rating* (0–10) as to how well it satisfies the want. If the alternative fulfills all possible aspects of a want, it would receive a rating of 10. On the other hand, if it only partially fulfills the want, it might receive a 4 or 5 rating. Multiply the weight of the want by the rating to arrive at a score for the want (weight × rating = score). For Want 1 in the table below, the weight is 8 (it is a fairly important want) and Alternative A does not do a very good job of filling this want so it is given a rating of 3. The score for Alternative A for Want 1 is 8 × 3 = 24. Do this evaluation for every want and add up the scores for each alternative, e.g., 59 for Alternative A and 65 for Alternative C. The alternative with the highest total score is your tentative first choice, which in this example is Alternative C.

Decision Analysis

Decision Statement: _____

_____

Alternative Solution:		A		C	
Musts	1. 2.	Go Go		Go Go	
Wants	Weight	Rating	Score	Rating	Score
1.	8	3	24	5	40
2.	5	7	35	5	25
		Total A = 59		Total C = 65	

While identifying weights and scoring may at first seem somewhat subjective, it is an extremely effective technique for those who can dissociate themselves from their personal biases and arrive at a logical evaluation of each alternative. If the alternative you "feel" should be the proper choice turns out to have a lower score than the tentative first choice, then you should re-examine the weight you have given to each want. Analyze your instincts to better understand which *wants* are really important to you. After this rescoring, if your favorite alternative still scores lower than the others, perhaps your "gut feeling" may be incorrect.

ADVERSE CONSEQUENCES OF THE ALTERNATIVE SOLUTIONS

The last step in the decision process is to explore the risks associated with each alternative solution. We take the top-scoring alternative solutions and brainstorm to make a list of all the things that could possibly go wrong if we were to choose that alternative solution. We then try to evaluate the *probability* (0–10) that the adverse consequence could occur and the *seriousness* (0–10) of this consequence *if* it were to occur. The product of these two numbers can be thought of as the *threat* to the success of the mission (*probability* × *seriousness* = *threat*). It is important not to let the numerical scores in the decision table obscure the seriousness of an adverse consequence. In some cases, the second-highest scoring alternative may be selected because the adverse consequences of selecting the highest scoring alternatives are too threatening.

An alternative to giving a numerical score would be to use the same evaluation system we used in the situation analysis. That is, we rate each adverse consequence as a low degree of concern (L), a moderate degree of concern (M), or a high degree of concern (H).

Adverse Consequences

Adverse Consequences	A. Probability of Occurrence	B. Seriousness If It Occurs	Threat
Alternative A 1. 2. 3.	0–10 or (L, M, H)	(0–10) or (L, M, H)	A × B
Alternative C 1. 2. 3.			

Decision Analysis:
Popara Enterprises Picks a New Market

Popara Enterprises, a multinational consumer goods company, will be shutting down a pulp-making operation over the next 12 to 16 months because demand for pulp has significantly decreased in recent years and is expected to continue to drop. Employees at Popara affected by the shutdown will be redeployed within the company to other manufacturing operations.

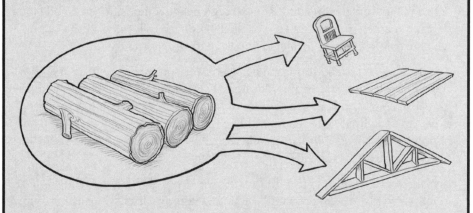

The greatest potential impact will be on the local wood businesses that supply the plant. About 15 logging companies and five sawmills located within 100 miles of the current pulp operation produce 30 tons of logs and chips each day. Popara Enterprises wants to work with its wood suppliers to develop new markets. The ideal solution would create new business as well as expand existing businesses to utilize all logs and chips currently produced. Timing is a critical factor. The sawmills and logging operations have payrolls as well as equipment financing costs to meet.

Decision Statement: Chose the best new market to redeploy the log and chip suppliers to use 30 tons of wood a day, cost under $5 million, and be ready within six months.

There are three options being considered for new markets: furniture, flooring, and house frames. All three have manufacturing startup costs under $5 million and can be ready in six months. The house-frame and flooring operations would use the required 30 tons of wood per day, but the furniture option would not, because producing furniture requires significant handwork beyond logging and sawing, leading to a slow production and less daily demand.

Continues

For the three options under consideration, tax relief would be the largest for the house frames, as local builders have lobbied for tax help for construction-related businesses. However, the tax relief in all cases would be fairly minimal.

House-frame workers need one month of training, furniture workers need three months, and flooring workers need roughly six months.

Competition is stiff in the furniture and house-frame markets where established brands and manufacturers in the region are well positioned. While the flooring market is competitive, no other producer is located in the region, which would give a new local merchant a market opportunity.

The flooring and house-frame industry look very promising, with home sales starting to steadily rise again over the past couple of years. Flooring sales have also benefited from strong sales of home improvement markets, a trend expected to continue into the future. The furniture market is good as well but is dominated by established brands. A new furniture operation would be likely to have a hard time gaining market share.

The three **musts** are (1) use 30 tons of wood per day, (2) be ready in six months, and (3) cost under $5 million. From the above discussion the furniture option will not be able to consume the minimum tonnage of wood required and is a *no go*, thereby excluding it from further consideration. The flooring and house-framing options meet all the musts.

Now we address the **wants** by arranging them in order of increasing importance, which in this case is tax relief, training, competition, and demand. Next we assign an appropriate **weight** to each. Tax relief, while desirable, is a minor consideration so we will give it a weight of 3. Training is deemed twice as important as tax relief so it receives a weight of 6. The lack of competition is slightly more important than training, so it is weighted 7. Finally, product demand is the most important so it's assigned a weight of 9.

The next step is to assign a **rating** to each want to determine how well the remaining options satisfy all the wants. The tax relief for house frames is larger than for flooring so we rate it an 8 for house frames and rate it a 4 for flooring. Training to make interlocking hardwood floor pieces takes longer than that to learn to make trusses for housing, so we rate these 4 and 7, respectively. There is not a local flooring competitor so we assign a rating of 7 for lack of competition and give house frames a rating of 5 because there are other builders in the area. There is a great demand for wood flooring both from existing homes as well as new construction, so flooring receives a rating of 9, while house frames, which rely solely on new construction, receives a rating of 5.

Alternative Solution		Furniture		Flooring		House Frames	
Musts 1. Use 30 tons of hardwood per day.		No Go		Go		Go	
2. Be ready in six months.		Go		Go		Go	
3. Manufacturing costs should be under $5 million.		Go		Go		Go	
Wants	**Weight**	**Rating**	**Score**	**Rating**	**Score**	**Rating**	**Score**
1. Product demand	9			9	81	5	45
2. Lack of competition	7			7	49	5	35
3. Minimal training	6	No Go		4	24	7	42
4. Tax relief	3			4	12	8	24
				Total = 166		Total = 146	

Summarizing the results from the KTDA table, we see that the best decision for the company is to create a wood flooring facility. We now need to carry out an adverse consequences analysis on our top two possibilities to learn if there is anything that would cause us to choose the second-highest score.

Adverse Consequences

We begin brainstorming the major things that could go wrong if they were to choose either of these solutions and then assess the probability (P: 1–10) that it could go wrong and how serious (S: 1–10) the consequence would be if that event did occur. The threat (T) to the choice of each alternative solution is then calculated as the product: $T = P \times S$.

For the flooring market, there is the threat of new market entrants; of particular concern would be a new competitor making flooring out of a different, possibly cheaper, material than wood. Synthetic wood can be used and could offer durability and cost benefits over hardwood floors. However, most premium buyers still prefer real wood.

Continues

Additionally, both need to be cautious of the bargaining power of suppliers. Construction companies often work with only one supplier and get large discounts for doing so. Discounts could decrease profit margins on house frames and possibly flooring too. Flooring can be sold through retail outlets, lessening the threat, but house frames cannot.

The housing-frame market is very dependent on the new housing market. If interest rates should increase and cause a decrease in the start of new housing, the need for new frames would be diminished.

Adverse Consequences	Severity (S)	Probability (P)	Threat (T)
Flooring —Product made cheaper from different material	6	6	36
—Bargaining power of customers	4	2	8
Total			44
House Frames —Interest rates cause housing market to falter	6	2	12
—Bargaining power of customers	4	7	28
Total			40

The final scores of the adverse consequences are sufficiently close so there is no reason to change our initial decision. The best decision is still creating a wood flooring facility.

Based on a real-life industrial situation.

To gain further facility with the KTDA, let's do one more real-life example to illustrate the DA procedure to choose among alternatives.

Choosing a Paint Gun

A new auto manufacturing plant is to be built, and you have been asked to choose the electrostatic paint spray gun to be used on the assembly line. The industry standard gun is Paint Right. While experience has shown that Paint Right performs adequately, its manufacturer is located in Europe, making service slow and difficult. The paint gun must provide a precise flow rate of paint to avoid both excessive waste from high flows as well as longer application times when low flow rates are required. The guns must also have a uniform spray droplet size so that the paint is applied evenly and produces a nice finished appearance. The desirable traits of the gun are durability, ease of service after a malfunction, low cost, and familiarity to the operators. While the European company, Paint Right, currently has the market share, its price is significantly inflated. Two American companies are eager to enter the market with their products: New Spray and Gun Ho.

Decision Statement: Choose an electrostatic paint spray gun. The paint guns available are Paint Right, New Spray, and Gun Ho.

Course of Action: The first step is to break down the important qualities of paint guns, and to decide what you must have and what you want to have. From your experience and discussions with other paint personnel, you determine that you have two musts: (1) adequate control over paint flow rate, and (2) acceptable paint appearance. Also, you identify four wants: (1) easy service, (2) low cost, (3) long-term durability, and (4) familiarity to plant personnel.

We first determine if each of the alternatives satisfy the musts. Plant records show that Paint Right is able to meet both musts. Next, laboratory experiments were performed with New Spray and Gun Ho to determine whether each of them is also able to meet both musts. The results showed that New Spray met standards but Gun Ho could not control the flow of paint at the level required, and thus it was dropped from consideration.

The four wants are then weighted and ratings are assigned for each gun that satisfies all the musts. The first want, ease of service, is an important one and receives a weight of 7. Because the parent company, Paint Right, is located in Europe, it is difficult to get rapid service so the rating for this want is 2, giving a score of $7 \times 2 = 14$. Durability is of moderate importance and receives a weight of 6 and Paint Right's gun is very durable and receives a rating of 8, giving a score of 48. The familiarity and low-cost wants are of lesser importance so they only receive a weight of 4. Because the plant personnel are currently using the Paint Right gun, it receives a rating of 9, giving a score of $4 \times 9 = 36$. We continue in this manner until the complete table is filled out as shown below.

Continues

Solution

MUSTS		Paint Right		New Spray		Gun Ho	
Adequate flow control		Go		Go		No Go	
Acceptable appearance		Go		Go		Go	
WANTS	Weight	Rating	Score	Rating	Score	Rating	Score
Easy service	7	2	14	9	63	No Go	
Durability	6	8	48	6	36		
Low cost	4	3	12	7	28		
Familiarity	4	9	36	2	8		
Total			110		135		

As can be seen from the solution in the table above, we have a clear winner of 135 points, New Spray's electrostatic spray paint gun. However, the company that is building the new plant still wants to tighten up all the loose ends and be sure that it has made the right choice before installing the new paint gun. In order to do so, it performed an adverse consequences analysis.

Adverse Consequences

Additional laboratory experiments were conducted with the two paint guns to determine the probability and seriousness of possible problems. Both paint guns could experience mechanical issues over time, affecting the performance and quality of the painting job. The probability of a mechanical issue, such as paint leaking from the container in the gun, did occur frequently when tests were performed with New Spray's paint gun ($P = 7$). The seriousness of this issue is low as there is no effect on the coating and it can easily be solved by refilling. The seriousness receives a rating of 2 for the New Spray gun. Paint Right's paint gun did not show any symptoms of leakage in extensive testing, giving a probability of 1. As the experiments progressed, the technicians noticed a slight change in the coating thickness for Paint Right's gun in comparison to the standard required thickness of paint, and they did not see a change in New Spray's results. Laboratory tests yielded that the probability of this defect occurring turned out to be 3 for Paint Right. This defect is considered to be very serious, because it will affect the appearance of the product.

Alternative	Probability of Occurrence (0–10)	Seriousness If It Occurs (0–10)	Threat (T = P × S)
Paint Right —Paint leaking —Nonuniform coating	1 3	2 9	2 27
Total			29
New Spray —Paint leaking —Nonuniform coating	7 1	2 9	14 9
Total			23

After applying an adverse consequences analysis, the scores of the analysis showed no reason to select the second-highest score. Consequently, New Spray's electrostatic spray paint gun was chosen over Paint Right's model for the new plant.

Based on a real industrial example.

Now, after using heuristic in industry settings, we move to an example applying the decision analysis to a personal life decision that many graduating students face.

Decision Analysis Example: Choosing Your First Job

Several years ago, a graduating senior from the University of Michigan used K.T. decision analysis to help him decide which industrial job offer he should accept. John had a number of constraints that needed to be met. Specifically, his fiancée (now his wife) was also graduating in chemical engineering at the same time, and both wanted to remain reasonably close to their hometown in Michigan. In addition, as a part of a dual-career family, John needed a guarantee that the company would not transfer him. After interviewing with a number of companies, he narrowed his choices to three companies: Dow Corning, ChemaCo, and TrueOil.

The first thing John did was to identify the musts that each offer had to satisfy; these criteria are shown in the KTDA table. Upon evaluating each company to learn whether it satisfied all the musts, he found that TrueOil did not satisfy the "no transfer" must. Consequently, John eliminated this company from further consideration. Next, he identified the wants and assigned a weight to

Continues

each criterion. The remaining two companies were then evaluated against each want and a total score was obtained for each company. Dow Corning scored 313 points, and ChemaCo scored 263 points; the apparent best choice was Dow Corning.

K.T. DECISION ANALYSIS: JOB OFFER

Objectives		Dow Corning			ChemaCo			TrueOil	
Musts									
In Midwest		Midland, Michigan	Go		Toledo, Ohio	Go		Detroit, Michigan	Go
Located within 40 miles of spouse's position		Another major company also in Midland	Go		Industrialized Northern Ohio	Go		Major metropolitan area	Go
No-transfer policy		Major plant in Midland	Go		Major plant in Toledo	Go		Must transfer	No Go
Wants	**Weight**		**Rating**	**Score**		**Rating**	**Score**		
Plant safety	10	Good (silicone)	7	70	Mainly oil derivatives (okay)	5	50		
Type of position	9	Process engineer	9	81	Pilot plant design and operation	10	90		
Salary and benefits	7	Good	6	42	Very good	7	49		
Near hometown (Traverse City, Michigan)	6	150 miles	10	60	400 miles	5	30		
Large company	4	Medium size	6	24	Small size	3	12		
Encourage advanced degree	4	Very positive	9	36	Positive	8	32		
Total				313			263		

Before making the final decision, John needed to evaluate the adverse consequences of his first and second choices. The results of the adverse consequence analysis are shown in the following table. The adverse consequences analysis ranked both choices in the same order as before; as a result of this analysis, the apparent first choice was confirmed as the final choice.

Alternative	Probability (P)	Severity (S)	Threat (P × S)
Dow Corning			
Midland is not very exciting	6	3	18
High rent	4	6	24
Total			42
ChemaCo			
Toledo is not very exciting	6	8	48
High rent	5	6	30
Total			78

Both John and his wife are working at Dow Corning in Midland, Michigan. (Only the names of the other companies were changed in this real-life example.)

Cautions

The assigning of weights and scores is indeed very subjective. One could easily abuse this decision-making process by giving higher weights/scores to a predetermined favored project. Such a biased weighting would easily skew the numbers and sabotage the decision-making process. The user is urged to refer to Kepner and Tregoe's book to become aware of certain danger signals that guarantee acceptance of a certain alternative and that blackball all others. This biasing could result from "loaded" want objectives, listing too many unimportant details, which obscure the analysis, or a faulty perception of which objectives can guarantee success. Consequently, it is very important to keep an open mind when making your evaluation. Look for fallacies in logic (see Chapter 3) throughout the decision-making process. Causal oversimplifications and Hasty generalizations are particularly common while making KTDA tables.

Missing Information

The most difficult decisions are those where you don't have all the necessary information available upon which to base a decision. Under these conditions it could be helpful after you have prepared a KTDA table to look at the extremes of the missing information and to perform a "What if …?" analysis. For example, just suppose in the job offer scenario that Dow Corning had not yet decided the type of position John would have with the company. John could assume the best case (his desired position of process engineer), which he would rate at 9, and the worst case in his opinion (e.g., traveling sales representative on the road full time), which would receive a low rating of 1. With this assumption, the total score for Dow Corning would drop to 241, which is now below the score of ChemaCo. We see that this want requires a key piece of information and that John must obtain more information from Dow Corning before he can accept that offer. If Dow Corning could not tell John which type of job he would have, it would at least be able to tell him which type of job he might *not* have (e.g., traveling sales representative). If it cannot do the latter, John could have been "forced" to choose ChemaCo. On the other hand, if all other factors are positive, John could decide to *take a risk* and choose Dow Corning with the chance he will be able to secure the desired position upon hiring or shortly after being hired. Epilogue: John has had a long and successful career with Dow Corning.

K.T. POTENTIAL PROBLEM ANALYSIS

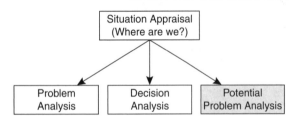

Having made a decision, you want to plan to ensure its success. We need to think about the future and consider what could go wrong and make plans to avoid these events if at all possible. To aid us in our planning, Kepner and Tregoe have suggested an algorithm that can be applied not only for ensuring the success of the decision but also for analyzing problems involving safety. The K.T. *potential problem analysis* (KTPPA) approach can decrease the possibility of a disastrous outcome. As with the other K.T. approaches, a table is constructed: The KTPPA table delineates the potential problems and suggests possible causes, preventive actions, and contingent actions.

K.T. Potential Problem Analysis

Potential Problem	Possible Causes	Preventive Action	Contingent Actions
A.	1. 2.		
B.	1. 2.		

In analyzing potential problems, identify how serious each problem would be if it were to occur and how probable it is to occur. Would the problem be fatal to the success of the decision (a must), would it merely hurt the success of the decision (a want), or would the problem only be annoying? First, brainstorm (see Chapter 7) to identify all the *potential problems* that could happen and list them in the KTPPA table. Be especially alert for potential problems when (1) deadlines are tight, (2) you are trying something new, complex, or unfamiliar, (3) you are trying to assign responsibility, and (4) you are following a critical sequence. Next, brainstorm and list all the *possible causes* that could bring about each problem. Next, develop *preventive actions* for each cause. Finally, develop a *contingent action* (last resort) to be undertaken if your preventive action fails to prevent the problem from occurring. Establish early warning signs to trigger the contingency plan. Do not, however, proceed with contingency plans rather than focusing on preventive actions.

K.T. Potential Problem Analysis: Jet-Lagged John

John is a safety engineer for Orange Manufacturing Corporation, living and working in Los Angeles. He is scheduled to fly to South Africa next week to perform safety inspections of his company's plants and give a lecture on how these industrial plants can make changes in their facilities to create a safer environment for their employees. Engineers from all around South Africa will attend his lecture.

John has only traveled occasionally for work and is not used to long flights. He has a weak stomach, and he was very tired and disoriented due to jet lag the only other time he was overseas. There is a nine-hour time difference between L.A. and Johannesburg in South Africa, where the conference is to be held. John was not able to get connecting flights that will get him to Johannesburg in less than 24 hours of travel time. John is worried about traveling such a long distance because of jet lag from the long journey and how it will affect him: possible motion sickness, dehydration, and cramps during the flights. In addition, John is up for promotion and needs to make sure he is both mentally and physically alert in order to give a stellar presentation.

John, being a well-educated engineer, decided to create a KTPPA table to make sure he is prepared for his trip and the worst-case scenarios. The complete table is presented below.

Potential Problem Analysis Table

Problem	Possible Cause	Preventive Action	Contingency Plans
1. Arrive late	• Flight delayed • Missed connection	• Leave a day early • Allow for added time between connections	• Video conference and inspect changes later
2. Dehydration	• Low cabin pressure	• Drink water and juices frequently • Use nasal spray • Use skin moisturizer	• Continue to drink fluids upon arrival • See a medical doctor upon arrival
3. Muscle cramps	• Sitting for long periods of time	• Don't stay seated for long periods of time; get up every two hours and walk around the plane • Do light exercises while seated	• Massage cramps
4. Disorientation	• Jet lag	• Arrive one or two days early • Get a good night's sleep before flight • Sleep on plane • Minimize alcohol and coffee consumption • Exercise on arrival	• Take sleeping pill the night before presentation

Continues

Potential Problem Analysis Table *(continued)*

Problem	Possible Cause	Preventive Action	Contingency Plans
5. Not feeling well	• Motion sickness	• Request window seat • Schedule flight on large (jumbo) aircraft • Bring medication	• Make sure air sick bag is available
	• Upset stomach	• Avoid overeating • Bring over-the-counter medication	• See medical doctor upon arrival
6. Poor presenta-tion	• Not prepared	• Have a dry run of the presentation • Brainstorm anticipated questions	• Fully respond to questions later
	• Computer damaged or lost	• Don't check computer: carry on plane	• Make arrangements to have a compatible computer available for presentation
	• Lost CD/DVD	• Make multiple copies and carry in different places	• Send a copy of the presentation on a CD ahead by UPS or FedEx or, if possible, attach to an email message, or place it in the cloud as backup

As an example of potential problem analysis, we will use a situation of buying a used car.

First, we brainstorm *all* **the serious potential problems** that could exist with a used car.

1. Concealed damage
2. Wheels not aligned
3. Suspension problems
4. Leaking fluids
5. Odometer incorrect

Next we list all **the possible causes** for each potential problem. Let's look at possible causes of *concealed damage*.

- Car was in a flood
- Car was in an accident

Next we list **the preventive actions** we could take so that we could determine if the *car was in a flood.*

- Take a deep whiff inside the car and trunk. Does it smell moldy?
- Look for rust in the spare tire well.

Finally we list **the contingent action** we would take if *the car was in a flood:*

- Don't buy the car.

The following table gives the complete K.T. potential problem analysis for buying a used car.

KTPPA: Lemon Aid—Buying a Used Car, Not a Lemon

Potential Problem	Possible Causes	Preventive Action	Contingency Plan
1. Concealed damage. Body condition is not what it appears to be.	Car was in a flood, window/trunk leak	Take a deep whiff inside the car and trunk. Does it smell moldy? Look for rust in the spare tire well.	Don't buy the car.
	Car in an accident or rusted out	Use a magnet along the rocker panels, wheel wells, and doors to check for painted plastic to which the magnet won't stick.	Offer a much lower price.

Continues

KTPPA: Lemon Aid—Buying a Used Car, Not a Lemon (Continued)

Potential Problem	Possible Causes	Preventive Action	Contingency Plan
		Look under insulation on the doors and trunk for signs the car was a different color.	
2. Buying a car that has improperly aligned front and back wheels.	Car was in an accident	Pour water on dry pavement and drive through to determine if the front and rear wheel tracks follow the same path or are several inches off.	Don't buy the car.
3. Car has suspension problems.	Hard use, poor maintenance	Check tire treads for peaks and valleys along the outer edges.	Require suspension to be fixed before buying.
4. Leaking fluids.	Poor maintenance	Look under the hood and on the ground for signs of leaking fluids.	Require seals be replaced before buying.
5. Odometer not correct.	Tampered with or broken	Check windows and bumpers for decals or signs of removed decals indicating a lot of traveling. Look for excessive wear on the accelerator and brake pedals. Check the title.	Offer a much lower price.
6. Car ready to fall apart.	Car not maintained during previous ownership	Check fluid levels (oil, coolant, transmission, brake). Check to see if battery terminals are covered with sludge. Check for cheap replacement of oil filters, battery, etc.	Don't buy the car.

Adapted from "Lemon Aid: To Buy, or Not to Buy, That's the Question," Hugh Hart, *Chicago Tribune*, September 20, 1992.

SUMMARY

One of the unique features of each of the K.T. strategies is the way they are able to display the data. In each case, with the KTSA, KTPA, KTDA, and KTPPA, you fill out a table and then analyze the data in that table in order to reach a decision.

SITUATION APPRAISAL

Problems	Timing (H, M, L)	Trend (H, M, L)	Impact (H, M, L)	Next Process (KTPA, KTDA, KTPPA)
1. 2. 3.				

PROBLEM ANALYSIS

	Is	Is Not	Distinction	Probable Cause
What				
Where				
When				
Extent				

DECISION ANALYSIS

Alternative Solution		A		B		C	
Musts	1. 2.	Go Go		Go No Go		Go Go	
Wants	Weight	Rating	Score	Rating	Score	Rating	Score
1. 2.					No Go		
		Total A =		Total B =		Total C =	

POTENTIAL PROBLEM ANALYSIS

Potential Problems	Possible Causes	Preventative Actions	Contingency Plan
A.	1. 2.		
B.	1. 2.		

Copyright Kepner–Tregoe, Inc., 1994. Reprinted with permission.

WEB-SITE MATERIAL (WWW.UMICH.EDU/~SCPS)

* **Learning Resources**
  Summary Notes—More Examples of Problem Statements

* **Professional Reference Shelf**
  The Pareto Analysis and Diagram

REFERENCES

1. Kepner, C. H., and B. B. Tregoe, *The Rational Manager*, 2nd ed., Kepner-Tregoe, Inc., Princeton, NJ, 1976.

2. Kepner, C. H., and B. B. Tregoe, *The New Rational Manager*, Princeton Research Press, Princeton, NJ, 1981.

3. Woods, D. L., A *Strategy for Problem Solving*, 3rd ed., Department of Chemical Engineering, McMaster University, Hamilton, Ontario, 1985; *Chemical Engineering and Education*, p. 132, Summer 1979; *AIChE Symposium Series, 79* (228), 1983.

EXERCISES

Situation Analysis

8.1. "The Exxon Valdez": It is 12:45 A.M. on March 24, 1989; you have just been alerted that the Exxon Valdez tanker has run aground on the Bligh Reef and is spilling oil at an enormous rate. By the time you arrive at the spill, six million gallons of oil have been lost and the oil slick extends well over a square mile.

A meeting with the emergency response team is called. At the meeting it is suggested that a second tanker be dispatched to remove the remaining oil from the Exxon Valdez. However, the number of damaged compartments from which oil is leaking is not known at this time and there is concern that if the tanker slips off the reef, it could capsize if the oil is only removed from the compartments on the damaged side.

The use of chemical dispersants (i.e., soap-like substances), which would break up the oil into drops and cause it to sink, is suggested. However, it is not known if there is sufficient chemical available for a spill of this magnitude. The marine biologist at the meeting objected to the use of dispersants, stating that once these chemicals are in the water, they would be taken up by the fish and thus be extremely detrimental to the fishing industry.

The use of floatable booms to surround and contain the oil also brought about a heated discussion. Because of the spill size, there is not enough boom material even to begin to surround the slick. The Alaskan governor's office says the available material should be used to surround the shore of a small village on a nearby island. The Coast Guard argues that the slick is not moving in that direction and that the booms should be used to contain or channel the slick movement in the fjord. The Department of Wildlife says the first priority is the four fisheries that must be protected by the booms or the fishing industry will be depressed for years, perhaps generations to come. A related issue is that millions of fish were scheduled to be released from the fisheries into the oil-contaminated fjord two weeks from now. Other suggestions as to where to place the boom material were also put forth at the meeting.

Carry out a K.T. situation analysis on the Exxon Valdez spill as discussed above.

8.2. "The Long Commute": The Adams family of four lives east of Los Angeles in a middle-class community. Tom Adams's commute to work is 45 miles each way to downtown L.A. and he is not in a car- or vanpool. He has been thinking about changing to a job closer to his home but has been working for over a year on a project that, if successfully completed, could lead to a major promotion. Unfortunately, there is a major defect in the product, which has yet to be located and corrected. Tom must solve the problem in the very near future because the delivery date promised to potential customers is a month away.

Tom's financial security is heavily dependent on this promotion because of rising costs at home. Both children need braces for their teeth, he is in need of a new car (it broke down twice on the freeway this past fall), the house is in need of painting, and there is a water leak in the basement that he has not been able to repair.

Sarah, Tom's wife, a mechanical engineer, has been considering getting a part-time job, but there are no engineering jobs available in the community. Full-time positions are available in northern L.A., but this would pose major problems with respect to chauffeuring and managing the children. There are a couple of day-care centers in the community, but rumor has it they are very substandard. In addition, last year, their son Alex was accepted as a new student by the premier piano teacher in the area and there is no public transportation from their home to his studio. Melissa is very sad at the thought of giving up her YMCA swimming team and her Girl Scout troop, which both meet after school.

Carry out a K.T. situation analysis on the Adams family's predicaments.

8.3. "First Day on the Job … Trial by Fire": Sara Brown has just become manager of Brennan's Office Supply Store, one of 10 in the Midwest. Sara's store, which is located in the downtown area on a busy street, has an inventory of over one million dollars and over 20000 square feet of floor space. On her first day of work, Sara is inundated with problems. A very expensive custom-ordered desk that was delivered last week received a number of scratches during unpacking, and the stockroom manager wants to know what he should do. She just discovered that the store has not yet paid the utility bills that were due at the end of last month, and she realizes that the store has been habitually late in paying its bills. The accounts receivable department tells her that it has had an abnormally high number of delinquent accounts over the past few months, and it wants to know what action should be taken. There is a large pile of boxes in the storeroom from last week that have yet to be opened and inventoried. The impression she has been getting all morning from the 30 employees is that they are all unhappy and dislike working at the store. To top things off, shortly after lunch, a large delivery truck pulls up to the front of the store and double-parks, blocking traffic. The driver comes into the store and announces that he has a shipment of 20 new executive desks. Where does Sara want them placed? The employees tell Sara that this shipment was not due until next week and there isn't any place to put them right now. Outside she hears horns of the angry drivers as the traffic jam grows.

Carry out a K.T. situation appraisal, based on the information below to decide what Sara should do.

A. While boxes on the floor may be an eyesore and awkward to step around, it is not necessary we do anything about them immediately. The situation is not getting worse by having them there, and the importance of not having them opened and the contents shelved is minimal. The next step is to decide who is to open the boxes and when to do it.

B. What to do about the 20 new desks has to be decided immediately because not accepting or accepting and storing such a large order would have major impacts on business. A traffic jam is beginning to form and is getting worse while Sara is deciding what to do.

C. The employee morale needs to be addressed in the very near future. It is believed that lack of care and sloppiness were factors in damaging the custom-ordered desk. The morale, while low, could get worse. We don't know why the morale is low.

D. Sara needs to pay the utility bills fairly soon or the electrical power to the store could be shut off, which would stop all functions of the business. Sara needs to find out why the money due to the store has not been paid.

E. Nothing needs to be done with the scratched desk immediately, but we do need to decide what to do in the not-too-distant future, and we also need to plan how to unpack the desks and other items more carefully.

8.4. Make up a situation similar to Exercises 8.1, 8.2, or 8.3, and carry out a K.T. situation analysis.

Decision Analysis

8.5. "Decision Analysis: iCrime": The theft of smart electronic devices, such as iPads, iPhones, Blackberries, and Droids, has exponentially grown in recent years, as each of these devices is fetching up to $400 on the secondhand market. With such high theft numbers and many attempted robberies turning into violent crimes, police departments and customers have mounted pressure on manufacturers to bring theft deterrents to market. Ideally, these deterrents need to be obvious enough to the thieves so they do not even bother attempting to steal the devices.

You are a systems engineer for a major electronic device manufacturer and must decide, by applying a K.T. decision analysis, what features to create for your company's smartphones and tablets in order to deter thieves.

The company's leadership team held a brainstorming session and laid out three musts for the project:

A. The implemented solution must not be susceptible to thieves erasing the device and defeating the protection.

B. The solution should cause minimal interference with the normal usage of the device.

C. The implemented solution needs to work in all countries and on all networks.

The team also presented four possible solutions that might help solve this theft trend.

Solution Options

1. *Device registry.* When a user purchases an electronic device he receives a serial number to register the device in his name in an online device registry database. When a device is stolen and reported, the serial number registry system will deny service to a stolen device when the new owner is trying to make a call, download data, or sign up for a new service with the same serial number but in someone else's name. The estimated additional price that the company will charge for implementing such a solution would be $10 for each buyer.

2. *Fingerprint/eye scan.* By installing a fingerprint or eye-scan activation system, which requires the scan of an authorized user's fingerprint or iris to use the device, the thieves will not have any use for the stolen device. The additional cost would be up to $20 per user for these high-security devices.

3. *Remote lock.* A self-disablement feature that will lock and turn off the device, by the remote push of a button from the owner or by calling the phone and entering a code to disable it, would make the stolen device unusable for thieves and would decrease the secondhand market of stolen devices. The additional cost of implementing this new software on the device would cost each user about $5.

4. *Find my device.* Another way to reduce the crime wave is to install a free (i.e., $0) application in the device software that has GPS tracking capability, making people always aware of where their devices are and enabling them to help the police track down the criminals faster if their devices would be stolen by providing information about the locations of stolen devices.

The board also informed you that it is important to keep in mind that the solution should add as little cost to the device as possible (weight = 9). The feature should reduce the functionality of the stolen device after the device is stolen (weight = 5). Additionally, the solution should also be reversible, meaning that if you lose your device by accident and then find it again you want to be able to reuse it (weight = 8).

All solutions work everywhere as long as they have at least a WiFi connection, that is, if the WiFi connection is enabled on the device. One drawback with device registry is that it is mainly applied for monthly and yearly phone contracts, not SIM-card user devices, and SIMs can be easily replaced. The fingerprint/eye-scan feature is thought to be a login function every time the device is logged off or turned off.

Prepare a K.T. decision analysis table and an adverse consequences scenario to decide which of the proposed solutions the electronic manufacturing company should implement.

8.6. "Decision Analysis: Choosing the Right Cloud": With the pace of business today, sharing and organizing documents and files so that they are continuously available from anywhere is of great importance. When working on group projects, editing and developing new material for magazines, books, or newspapers can be very messy and confusing when different versions of different documents are sent back and forth before a final version is agreed upon. This confusion only adds to stress as deadlines quickly approach. Today, it is very common to have more than one computer, smartphone, and tablet, adding additional dimensions to the confusion of having multiple versions of multiple documents existing in multiple places.

One group that is experiencing this kind of confusion and lack of organization with keeping their documents updated is the writers at the *Chem-Daily* newspaper. The group has considered investing in a cloud service system, an online file sharing and storage service. Cloud service systems provide convenience by gathering all documents into one place and synchronizing them online with all devices and users. The group researched cloud service systems and found four possible candidates: Dropbox, SugarSync, Microsoft's SkyDrive, and Google Drive.

The first is Dropbox, the most popular and best-known cloud service. It is very simple to use. A folder is created on your device and all the files that are placed in that folder will automatically be synchronized to all other devices and users. The Dropbox service provides 2 GB of free storage and users can sign up for a 100-GB plan for $99/year. Another cloud service system with functions similar to Dropbox, again using folders, is SugarSync. Its advantage is that it does not just synchronize one specific folder. Instead, it backs up the existing structures of your folders on your device, avoiding the problem of forgetting to place the documents in a specific folder. SugarSync offers a 5-GB free storage and for an annual cost of $150, the 100-GB plan. The one cloud system that offered the most free storage space, 7 GB, was Microsoft's SkyDrive. This cloud system has the function of a cloud system combined with Microsoft's Office tools. This built-in feature allows the users to edit and write documents directly in the cloud. Microsoft's SkyDrive also provides a 100-GB plan but for a lower price of $50/year. Google Drive, another cloud system that has the same features as Microsoft's SkyDrive with built-in editing capabilities, offers a 5-GB free plan and a 100-GB plan for $60/year. However, Google Drive requires files and documents to be converted into Google's format in order to be usable in the cloud system. Finally, the group found that all cloud service systems are available to download on smart devices.

The group laid out some musts and wants for a cloud service system. The three musts are (1) the cloud system must work on both Macs and PCs, (2) it must have a large storage space (at least 5 GB), and (3) files need to be easily shared. For the wants, the group prefers that the cloud system be available on smart devices, such as iPhones and iPads, to be more effective and productive, on the road to and from the office (weight = 4). An important want the group has is that the cloud service system should provide the option to directly edit the files in the cloud system, through a built-in feature, without the need to first save it to the computer and then open the edit program to start working (weight = 7). A third want was that the cloud system should have a cheap annual price (below $100/year for the 100-GB plan) (weight = 9).

Help the group decide which cloud service system it should choose by applying a K.T. decision analysis.

8.7. "The Perfect Vacation": Lara and Tom Anderson are a couple, both 28, who want to take a summer vacation from their hectic schedule. They live in the suburbs of Chicago and commute 45 minutes each way on crowded trains to downtown Chicago and are eager to get away from the hustle and bustle and experience a change of scenery. They have set aside $2000 for the vacation, which is the most they can spend. They both love learning new things, seeing interesting sights, and eating good food. However, Tom is currently on a diet. Tom likes to do outdoor sporting activities, while Lara likes to visit art museums and is happy to sit outside and read. The main thing that both Lara and Tom want to do is come back refreshed. They have brainstormed and identified three vacation possibilities.

Lara tells Tom of three museums she would like to visit in New York. While the airfare is cheap, hotels are expensive and they would have to get a hotel outside the city and commute to stay within their budget. Tom says that would be okay as he enjoys theater and there are excellent restaurants there, but he also suggests northern Michigan, where he can golf, sail and visit Mackinac Island, which has a lot of rich history, fun restaurants, and hiking trails. Upper Michigan also has the Mackinac Bridge, world-class golf courses in Traverse City, and Sleeping Bear Sand Dunes, where you can hike and travel across the dunes in a dune buggy. Lara shows Tom a brochure from Cancun, Mexico, where they can spend a week in the sun sailing and lying on the beach for $895 per person, airfare, lodging, and meals included. Cancun is not known for its food and Tom doesn't like spicy things. She points out there are guided tours of the Mayan ruins only two hours away.

Lara and Tom agree to do a K.T. decision analysis on their options. Their musts are cost limit, new sights, and no stress. All vacation alternatives will cost under $2000, and Tom and Lara have never been to any of the alternatives. New York would be stressful because they feel that they would have the same stressful commute each day they do in Chicago and would not come back refreshed; however, the other options would not be stressful.

Tom and Lara narrowed their wants list down to three: nice views and scenery, food, and learning new things. The views of the clear Gulf water and sandy beaches in Cancun are wonderful and very appealing. The meeting of the Great Lakes (Lake Michigan and Lake Huron) and the Mackinac Bridge is a wonderful site as is Mackinac Island, which doesn't allow cars.

There are excellent restaurants in northern Michigan, such as Tapawingo, which is listed in Zagat. Tom does not like spicy food and the restaurants in Cancun, while good, are not at the same level as northern Michigan.

Learning is their most important want. In Cancun, learning the Mayan history would be quite rewarding. In northern Michigan, there is Fort Mackinac (from the Revolutionary War) and Fort Michilimackinac (from the French-Indian War in 1715): both are rich in history but not as interesting as the Mayan history.

Once the best tentative decision is reached from the K.T. decision analysis, Lara and Tom must consider the adverse consequences of each of the alternate solutions that is a "go."

Bad weather could be a factor in both of the final choices. The probability of bad weather in northern Michigan at this time of year is very low while it is hurricane season in the Gulf Coast and the probability of a hurricane is relatively high, but it might not hit Cancun. However, a hurricane hitting Cancun would be very serious if it were to occur. Crowds in northern Michigan could make it difficult to get into restaurants and museums and the probability of this occurring is moderate and the seriousness is mild if it were to occur. A guide strike in Cancun would cause them not to see the Mayan ruins, but the probability of occurrence is extremely low. However, a guide strike would be very serious if it did occur.

Carry out a decision analysis for Tom and Lara.

8.8. "Buying a Car": You have decided you can spend up to $12000 to buy a new car. Prepare a K.T. decision analysis table to decide which car to buy. Use your local newspaper to collect information about the various models, pricing, and options and then decide on your musts (e.g., air bag) and your wants (e.g., quadraphonic stereo, CD player). How would your decision be affected if you could spend only $9000? What about $18000?

8.9. "Choosing an Elective": You need one more three-hour nontechnical course to fulfill your degree requirements. Upon reviewing the course offerings, and the time you have available, you note the following options:

- Music 101   Music Appreciation, 2 hours
- Art 101   Art Appreciation, 3 hours

- History 201   U.S. History: Civil War to Present, 3 hours
- Art 203   Photography, 3 hours
- Geology 101   Introductory to Geology, 3 hours
- Music 205   Piano Performance, 2 hours

Music 101 involves a significant amount of time outside of class listening to classical music. The student reaction to the class has been mixed; some students learned what to listen for in a symphony, while others did not. The teacher for this class is knowledgeable but boring.

Art 101 has the students learning the names of the great masters and how to recognize their works. The lecturer is extremely boring and you must go to class to see the slides of the great art works. While the course write-up looks good, it misses the mark in developing a real appreciation of art. However, it is quite easy to get a relatively good course grade.

History 201 has an outstanding lecturer that makes history come alive. However, the lecturer is a hard grader and C is certainly the median grade. In addition, the outside reading and homework are enormous. While some students say the workload is equivalent to a five-hour course, most all say they learned a great deal from the course and plan to continue the interest in history they developed during this course.

Art 203 teaches the fundamentals of photography. However, equipment and film for the course are quite expensive. Most of the time spent on the course is outside of class looking for artistic shots. The instructor is very demanding and bases his grade on artistic ability. Some students say that no matter how hard you work, if you don't develop a "photographic eye," you might not pass the course.

Geology 101 has a moderately interesting lecturer and there is a normal level of homework assignments. There are two major out-of-town field trips that will require you to miss a total of one week of class during the term. The average grade is B and there is nothing conceptually difficult or memorable about the course.

Music 205 requires you to pass a tryout to be admitted. While you only spend half an hour a week with your professor, many, many hours of practice are required. You must have significant talent to get a C or better.

Prepare a K.T. decision analysis table to decide which course to enroll in.

8.10. "The Centralia Mine Fire": Centralia, Pennsylvania, a small community situated in the Appalachian mountain range, was once a prosperous coal-mining town. In 1962, in preparation for the approaching Memorial Day parade, the landfill of Centralia was set afire in order to eliminate odors, paper buildup, and rats. Unfortunately, the fire burned down into the passageways of the abandoned mine shafts under the town. Although repeated efforts were made to stop the blaze, the fire could not be put out. By 1980, after burning for 18 years, the fire had grown in size to nearly 200 acres, with no end in sight.

Mine fires are especially difficult situations because they are far below the surface of the earth, burn very hot (between 400°F and 1000°F), and give off both toxic and explosive gases, as well as large volumes of steam when the heat reaches the water table. Anthracite coal regions have very porous rock, and consequently, a significant amount of combustion gas can diffuse directly up through the ground and into people's homes. Subsidence, or shifting of the earth, is another serious condition arising from the fire. When the coal pillars supporting the ceilings of mines' passageways burn, large sections of earth may suddenly drop 20 or 30 feet into the ground.

Clearly, the Centralia mine fire has very serious surface impact and must be dealt with effectively. Several solutions to the mine fire are described below. Perform a K.T. decision analysis to decide which option is the most effective method to deal with the fire. Consider such issues as cost, relocation of the residents of Centralia, and potential success of extinguishing the fire.

Solution Options

   A. *Completely excavate the fire site*. Strip-mine the entire site to a depth of 435 feet, digging up all land in the fire's impact zone. This would require partial dismantling of Centralia and nearby Byrnesville for upward of 10 years, but available reclamation techniques could restore the countryside after this time. This method guarantees complete extinction at a cost of $200 million. This cost includes relocation of families, as well as the restorative process.

   B. *Build cut-off trenches*. Dig a trench to a depth of 435 feet, then fill it with a clay-based noncombustible material. Behind the trench, the fire burns unchecked but is contained by the barrier. Cost of implementation would be about $15 million per 1000 feet of trench, and total containment of the fire would require approximately 7000 feet of trench. Additionally, partial relocation of Centralia would be required for three years, costing about $5 million.

   C. *Flood the mines*. Pump 200 million gallons of water per year into the mine at a cost of $2 million annually for 20 years to extinguish the fire. Relocation of the townspeople is not necessary, but subsidence and steam output should be considered, as well as the environmental impact and trade-offs of the large quantities of acidic water produced by this technique.

   D. *Seal mine entrances to suffocate the fire*. Encase the entire area in concrete to seal all mine entrances, then allow the fire to suffocate due to lack of air. This would require short-term relocation of the town and its outlying areas, and suffocation itself would probably take a few years, owing to the large amount of air in the shafts and in the ground. Although this method has never been attempted, the cost is estimated to be about $100 million.

E. *Use fire-extinguishing agents*. Pump halons (gaseous fluorobromocarbons) into the mines to extinguish the blaze. The cost for this method would be on the order of $100 million. Relocation may be necessary.

F. *Do nothing*. Arrange a federally funded relocation of the entire area and allow the fire to burn unchecked. Approximately $50 million would be required to relocate the town.

This problem was developed by Greg Bennethum, A. Craig Bushman, Stephen George, and Pablo Hendler, University of Michigan, 1990.

8.11. "Solvent Body Wipe Problem": When painting a new automobile it is essential that the exterior surface is clean. Any impurities present when the paint is applied will severely degrade the quality of the paint finish. Consequently, line operators use solvent-moistened wipes to clean the surface of each vehicle body before it enters the paint booth.

Current body wipe operations at one of the Big Three auto companies can be summarized as follows. The operator soaks a new, dry body wipe (a white synthetic cloth about 1 foot square) in an open bath of solvent. There is a small amount of spillage and evaporative solvent loss from the bath. After wringing out any excess solvent, the operator wipes down the exterior of the vehicle body. The operator then disposes of the wipe in a drum and moistens a new wipe for the next vehicle. Because the wipes in the drum still contain significant amounts of solvent—a hazardous and possibly toxic substance—the drum is sealed when it becomes full and is sent to a hazardous waste landfill. The safety officer has expressed concern about the level of operator exposure to the toxic solvent fumes.

This cleaning operation must be applied to each vehicle body as it moves along a conveyor belt from the assembly shop into the paint booth. Depending on demand, the production rate may require painting up to 80 bodies per hour. When defects in the paint job are found, the body must be taken offline for spot repairs or complete repainting. If a large percentage of bodies have paint defects, then production of the entire plant can be diminished as repaint jobs back up at the paint shop.

Upper-level management has decided to open another plant that will also require body wipe operations. Due to more stringent environmental, health, and safety regulations for new plants, the liquid hazardous waste must be reduced without slowing production. If possible, management would also like to make other improvements to the current method of cleaning.

Problem: What is the best body wipe method to adopt in the new plant? Engineering proposed four alternative methods which are summarized below:

1. *Use each wipe on more than one vehicle*. This alternative requires less effort by the operators since they do not need to moisten a wipe as often. Less solvent is applied per vehicle. Not only are the additional operational costs of this alternative

negligible, but raw material costs, waste solvent, and waste wipes are also significantly decreased. However, there is a small chance that reusing wipes would leave impurities on the vehicle and cause the paint coat to crater. While a few vehicles with impurities would not slow production, many vehicles would.

2. *Recycle*. This option requires expensive additional equipment to wash the wipes and recover solvent. The amount of solvent applied to each vehicle would remain the same but consumption of virgin raw materials would be decreased moderately. Disposal of waste wipes and solvent would be decreased moderately as well.

   However, this process should be monitored closely as recycled wipes would likely deteriorate after a number of washings and leave lint on the vehicle, thereby causing paint defects.

3. *Incinerate used body wipes*. With this option, the energy value of the spent wipes can be partially recovered. In addition, the need for disposing of hazardous waste is totally eliminated. It is possible, but unlikely, that the ash will need to be sent to a landfill as nonhazardous waste. With incineration, there are no changes in the actual wiping process; the option is completely "end-of-pipe." Therefore, solvent usage will be unaffected. Incineration is moderately expensive.

4. *Use a closed-top bath*. The solvent bath would be covered with a lid that the operator would remove each time he soaks a rag. This option would cut down on solvent evaporative losses and spillage from the bath, both of which management desires. However, the amount of solvent applied to each vehicle would stay the same.

Management gives only moderate weight to the actual amount of hazardous waste diverted from the landfill and to decreasing raw material usage and minimizing operator effort. The degree of workplace safety hazards (an extremely important consideration) is influenced weakly by the amount of solvent stored on-site, which in turn is influenced by the amount of solvent applied to each vehicle. The additional operating cost of the new method would also be a major consideration. Finally, management wants the adverse consequences of the top two options to be addressed, especially the negative impacts on product quality—a very serious issue. Carry out a K.T. decision analysis to select a method to wipe the cars.

8.12. You need energy for an upcoming sports competition. You have the following candy bars available to choose from: Snickers, MilkyWay, Mars Bar, Heath Bar, and a granola bar. Which do you choose? Prepare a K.T. decision analysis table.

8.13. Prepare a K.T. decision analysis table on selecting an apartment to move into next year. Consult your local newspaper to learn of the alternatives available.

Potential Problem Analysis

8.14. "Sandy Beach": There was a minor oil spill on a small sandy resort beach. The CEO of the company causing the beach shoreline to be soiled with oil said, "Spare no expense, use the most costly method, steam cleaning, to remove the oil from the sand."

Carry out a K.T. potential problem analysis on the direction given by the CEO.

Adapted from *Chemtech*, August 1991, p. 481.

8.15. "Laboratory Safety": The procedure in a chemistry laboratory experiment called for the students to prepare a $1.0 \, dm^3$ aqueous solution of 30 g of sodium hydroxide. By mistake, the student used 30 g of sodium hydride dispersion, which reacted violently with water, evolving heat and hydrogen gas, which caught fire. The sodium hydride, which was available for a subsequent experiment, was a commercial product. The container bore a warning of the hazard of contact with water, but this warning was not visible from the side showing the name of the compound.

Carry out a potential problem analysis that, if followed, would have prevented this accident.

Adapted from *ICE Prevention Bulletin*, December 1991, 102, p. 7.

8.16. "Safety in the Plant": A reactor approximately 6 feet in diameter and 20 feet high in an ammonia plant had to be shut down to repair a malfunctioning nozzle. The nozzle could be repaired only by having a welder climb inside the reactor to carry out the repair. During welding, the oxygen concentration was regularly monitored. Four hours after the welding was completed, a technician entered the reactor to take pictures of the weld. The next day he was found dead in the reactor.

Prepare a potential problem analysis table that could have prevented this accident.

Adapted from *ICE Prevention Bulletin*, December 1991, 102, p. 27.

8.17. "New Chicken Sandwich": Burgermeister has been serving fast-food hamburgers for more than 20 years. To keep pace with the changing times and tastes, Burgermeister has been experimenting with new products in order to attract potential customers. Product development has recently designed a new Cajun chicken sandwich to be called Ragin' Cajun Chicken. The developers have spent almost nine months perfecting the recipe for this new product.

One of the developers got the idea for a new product while in New Orleans during last year's Mardi Gras. Product Development has suggested that the sandwich be placed on Burgermeister's menu immediately, in order to coincide with this year's Mardi Gras festivities. A majority of the time spent developing the Ragin' Cajun Chicken sandwich was dedicated to producing an acceptable sauce. Every recipe was tasted by the developers, who found early recipes for sauces to be too spicy. Finally, they agreed on the seventy-eighth recipe for sauce (Formula 78) as the best choice.

After converging on a sauce, the Development team focused on preparation aspects of the new sandwich. Several tests confirmed that the existing equipment in Burgermeister restaurants could not be used to prepare Ragin' Cajun Chicken. Instead, a new broiler would have to be installed in each of the 11000 Burgermeister restaurants, at a cost of over $3000 per unit. The new broiler would keep the chicken moist while cooking it, as well as killing any salmonella, the bacteria prevalent in chicken.

While testing cooking techniques for the new broiler, one of the developers became very ill. A trip to the hospital showed that the developer had food poisoning from salmonella. Tests determined that the source of the bacteria was a set of tongs that the developer used to handle both the raw and the cooked chicken.

Next, the Development team decided how the sandwich would be prepared. When the Ragin' Cajun Chicken sandwich was prepared using buns currently used for other Burgermeister sandwiches, the sandwich received a very low taste rating. After experimenting, researchers found that a Kaiser roll best complemented the sandwich. Early cost estimates showed that Kaiser rolls will cost twice as much as the buns used currently for hamburgers, and are fresh half as long.

You are an executive in charge of product development for Burgermeister. Based on the information above, perform a potential problem analysis, considering what could go wrong with the introduction of this new sandwich.

Developed in collaboration with Mike Szachta, University of Michigan, 1992.

8.18. Carry out a KTPPA for each of the following situations:

    A. A surprise birthday party.

    B. A camping trip in the mountains.

    C. The transportation of a giraffe from the Detroit Zoo to the Los Angeles Zoo.

    D. An upcoming laboratory experiment.

    E. The transport of nuclear waste from the reactor to the disposal site.

FURTHER READING

Keith, Lawrence A. "Report Results Right!" Parts 1 and 2, *Chemtech*, p. 351, June 1991, and p. 486, August 1991. Guidelines to help prevent drawing the wrong conclusions from your data.

Kepner, C. H., and B. B. Tregoe. *The New Rational Manager*. Princeton Research Press, Princeton, NJ, 1981. Many more worked examples of the K.T. strategy.

9 IMPLEMENTING THE SOLUTION

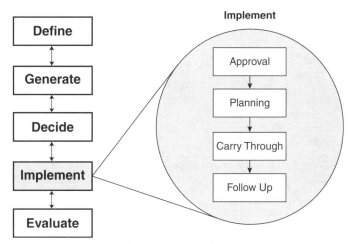

Implementing the Solution

Many people get stalled in the problem-solving process because they overanalyze things and never get around to acting. In this chapter, we present a number of techniques that will facilitate the **implementation process,** which can act as a guide in moving forward instead of getting stuck. The figure on page 230 identifies the phases of the implementation process.

Each of the phases of implementation will be discussed in this chapter and is applicable to both the problem-solving process and general project planning.

APPROVAL

Imagine we are involved in the problem-solving process on the job. We have identified the real problem, brainstormed and selected a solution, and are ready to implement it. We have developed both a project plan for our solution and a business plan to consider the financial effects of our project. Once we have developed our project and business plans to our satisfaction, we must get approval from our management or those to whom we report before we proceed any further. In the case of a small business or entrepreneurial project it might be a supervisor who can give the approval. However, for a major project at a large company it could be the CEO, president, or perhaps even the board of directors from whom we will need to receive the green light for the path forward. When seeking approval it is important to keep in mind that sometimes certain steps from the planning phase must be taken out of order. Presenting a well-constructed plan for implementation can alleviate anxiety for the person giving approval to a large complicated project. In these cases, simply go through the planning steps detailed in this chapter first as needed, then return to get approval to implement the plan.

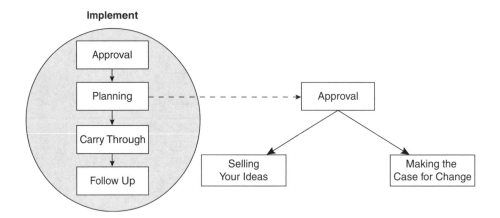

Selling Your Ideas

Many times it may be necessary *to sell your ideas* so that your organization will provide the necessary resources for you to complete your project successfully. This process may include the preparation of a document or presentation describing (1) what you want to do, (2) why you want to do it, (3) how you are going to do it, and (4) how your project will greatly benefit your organization and/or others. During your presentation to sell your ideas, remember these points:

- Avoid technical jargon—keep your presentation clear and to the point.
- Give your presentation in a logical and orderly manner.
- Be concise; avoid unnecessary, minute details.
- Anticipate questions and be prepared to respond to them.
- Be enthusiastic about your ideas—or no one else will be.

Lightspring/Shutterstock

The "Responding to Criticism" section in Chapter 2 can be a great resource when you encounter difficult people to deal with while selling your ideas during the approval process.

Making the Case for Change

Leaders and entrepreneurs in and out of the workplace will often be tasked with creating and implementing change in an organization. This can be a daunting task, especially when change is controversial, but nevertheless learning to successfully champion change is a very valuable skill. The following is a simple process that can be used to manage change.

Simple Process

First	*Articulate the case for change.* What is the change? Why do we need to make the change? Who will the change affect?
Second	Prepare a *vision* outlining what it will be like after the change.
Third	Identify the *skills* needed to make the change (communication, marketing, design, etc.).
Fourth	Define *incentives* for change (What do we get if we change? What is the benefit for the organization?).
Fifth	Identify the *resources* to implement the change (Can we afford it? Do we have the appropriate personnel to bring about the change? etc.).
Sixth	Have an *action plan* (timetables, Gantt charts, and critical path).

Source: Courtesy of Hank Kohlbrand, Dow Chemical Company (retired), and 2010 President of the American Institute of Chemical Engineers.

Completion of all six steps in progression will help lead to successful change. However, if one (or more) of them is left out, the results will be less than satisfactory. The following diagram lays out what happens when any singular step is missed.

Managing Complex Change

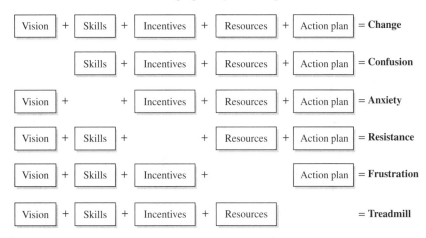

Adapted from Knoster, T. (1991). Presentation at TASH Conference, Washington DC (Adapted by Knoster from Enterprise Group Ltd.)

These attributes can be used as warning signs during the change process. Good managers will see these attributes arising and will work to make the necessary improvements. For example, if a manager sees the desired change is meeting resistance, he or she could create a new incentive or improve the current incentive for change.

Many times, companies and organizations decide from the top down to restructure or modify something to improve operations. After the decision has been made, the implementation of the change can become disruptive, waste time, and irritate those who are affected by the decision. The following is a recent case history.

25 Years or Less?

A professional organization wanted to recognize its members who had achieved a high level of accomplishment and who had also given a significant amount of time providing service to the organization. These members would be given a special membership rank called Fellow, a lapel pin, and a certificate of recognition. The required service activities included serving on and chairing a number of the organization's committees as well as other leadership roles such as serving as an officer of the organization. Initially, the recognition was only allowed to be given to an individual with a minimum of 25 years of professional practice after receiving a bachelor's degree. The executive branch of the organization decided to make more people eligible for this Fellow recognition and certificate by removing the restriction of 25 years of professional practice after receiving a bachelor's degree.

The removal of the 25-year requirement caused an uproar among the current Fellows who had already received this recognition. Upon hearing of this impending change, some of the current Fellows made comments such as "I had to wait 25 years, so why shouldn't they?", "We will lower the status of the recognition if we do this," "It's okay with me if they can meet all the criteria in less than 25 years," and "I will resign my membership if the organization implements this change." This pushback from the members was taking time away from the executive branch's other priorities.

The leadership clearly did not plan ahead to avoid these feelings of animosity from Fellows who had their certificates and therefore had a difficult time enacting the change. Following the algorithm for managing complex change would have looked like this:

- The leaders *articulate* that changing the rules for recognition is important because it would help the organization retain the most talented mid-career professionals. The award becomes truly about accomplishments, not longevity.

- They discuss their *vision* that the benefits of recognizing more high-quality young members will raise the status of the organization and help preserve it for the long term.

- The leaders identify a member with great communication *skills* to clearly explain and convince the older members that the quality of people being considered will keep the recognition level at a high level.

- The leaders define *incentives* for the change such as rewarding the new Fellows for their hard work for the organization and thereby encourage them to take on even greater levels of participation in the organization.

- The leaders identify what *resources* they will use to implement the change, such as using an organizational magazine to point out the reasons for the change.

- The leaders prepare an *action plan* that shows the period over which the transition will happen, such as limiting the number of less-than-25-year applications that can be submitted in the first year.

This example is just one of many that indicates the need for a set of guidelines for managing changes.

The following "case for change" example is from the 2012 London Olympic Planning Committee.

Changing Buses

Many sports fans come from all across the United Kingdom and Europe in buses to rugby and soccer matches in London. The buses from the various towns usually park in the stadium parking lot where the fans can leave their belongings, such as backpacks, on the bus and go inside the stadium for the match. After the event they return to the buses, which will take them directly back to their homes.

However, for the Olympics a different procedure has been suggested. The fans would have to transfer from their home bus to a London double decker city bus at one of several satellite locations on the circular M25 Motorway surrounding London. The proposed change could upset the fans and bus drivers if not handled properly. Make the case for change.

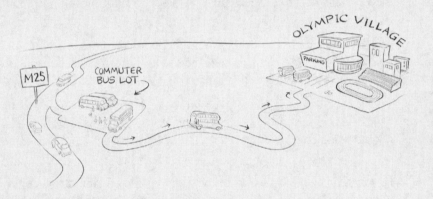

Articulate the Case for Change

What is the change?

The fans from other cities would transfer at satellite lots along M25 from their home bus to a London city bus to take them to the event.

Why does the change need to be made?

1. Danger of terrorism. The buses from other countries that would park next to a sporting event may contain explosives that may not be detected.

2. Non-London drivers could get the fans to the events late. The traffic pattern in London will be drastically changed and the non-London drivers of the travel buses could wind up getting lost with all the rerouting, whereas the London bus drivers would receive special training and routes to get fans to events efficiently and safely.

Vision of the Implemented Change

The visitor buses would be parked at the satellite locations and London buses would continually make round trips from the Olympic village to these satellite locations.

Skills Needed to Make the Change

Identify those individuals in charge of various components of the change and make sure they have the *skills* to prepare the schedules and manage the satellite locations. These individuals will need good organizational and communication skills to carry out such tasks as they communicate with the person in charge of the satellite parking, the liaison with the bus drivers outside of London, the London bus drivers, and the advertising agents in charge of letting visitors know what the change will be and helping them plan for it.

Advertise the Incentives for the Change

Prepare materials that publicize the reasons for the change, namely that it would prevent any vehicles rigged with explosive devices from parking next to a building hosting a sporting event. Promote the fact that shuttle parking will make for a safer Olympics, helping deter terrorist activities that could occur in the Olympic complex.

Identify Resources to Implement the Change

The number of London city buses will have to be increased to be sufficient to carry all the fans to the events on a specified day. Make sure there is sufficient parking space for the out-of-town buses as well as the necessary personnel to help direct the fans to the appropraiate buses to take them to the sporting event for which they have purchased a ticket.

Have the necessary funds to recruit paid and nonpaid volunteers to work the satellite lots and the lot at the Olympic village.

Make a Plan

Identify and communicate the different location at which the buses will be stationed in the satellite lots in order to take the fans to the different sporting events.

Register the buses' travel agencies three months in advance to secure a parking place in a satellite lot along the motorway.

Arrange for the billboards along the motorway and properly identify the various satellite lots and have signs directing unfamiliar drivers where to go and park.

Make sure the London buses will be appropriately marked so that after the event the fans can find a bus to take them to their correct satellite bus location.

Prepare timelines and Gantt and deployment charts for each phase of the operation (e.g., notify bus companies all over the United Kingdom and Europe that they will need to register).

PLANNING

Now that we have approval for the project, it is time to plan what to do, in what order to do it, and when to do it. The most important aspect of implementation is the **planning phase.** In this phase, we look at the resource allocations of time, personnel, and money; anticipate bottlenecks; identify milestones in the project; and identify and sketch the pathway through to the finished solution. A modified K.T. situation appraisal (see Chapter 8) will help to identify the critical elements of the solution and to prioritize them so that we can prepare a meaningful plan. Gantt charts, deployment charts, budgets, and critical path management[1] can all be used to effectively allocate time and resources. Finally, we proceed to identify what might go wrong and devise ways to prevent these events from occurring; that is, we use K.T. potential problem analysis (KTPPA; discussed in Chapter 8). In many industries, for example, market surveys are used as a part of the KTPPA to anticipate the possible success or failure of a product or process.

> *"If you don't know where you are going, you'll probably end up somewhere else."*
> —Yogi Berra

In the planning stage, we use resource allocation along both with K.T. situation appraisal and K.T. potential problem analysis as discussed in Chapter 8 and shown in the following figure.

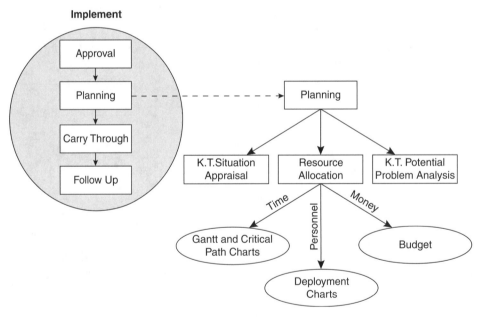

Components of the Planning Process

Resource Allocation

Having been presented with a problem, situation, or opportunity, we need to allocate our time and resources to the various steps wisely to bring about a successful solution. We can use a variety of techniques—the Gantt and deployment charts and strategies for budgeting both personnel and money—to arrive at an efficient and effective allocation. For individuals, proper time allocation and scheduling are important for success. High-functioning executives and students alike use detailed planners and calendars to keep track of important appointments and commitments.

Gantt Charts

One of the most popular ways to allocate specific blocks of time to the various tasks in a project is a **Gantt chart.** A Gantt chart is a bar graph that indicates when a specific task is to begin and how long it will take to complete that task.

As an example, suppose we have a time constraint of one year to solve the problem and we need to allocate time to each of the five building blocks of the problem-solving process. January, February, and March will be spent working on problem definition; April and May will be devoted to generating solutions. We suggest that time be allocated to evaluate our progress at four points along the way to check that all criteria have been fulfilled: (1) after completion of the definition of the problem, (2) after deciding the course of action, (3) during the implementation, and (4) at the end of the project.

TASK	MONTH											
	J	F	M	A	M	J	J	A	S	O	N	D
Problem Definition	▨	▨	▨									
Generate Solutions				▨	▨							
Decide Course of Action						▨	▨					
Implement								▨	▨	▨	▨	▨
Evaluate				▨			▨		▨			▨

A Gantt Chart

In the Gantt chart, note that at least 25% of the time has been devoted to the problem definition process, which includes the four steps of gathering information discussed in Chapter 4. Many—if not most—of the unfortunate consequences of the incorrectly defined problems discussed in Chapters 1 and 5 would not have occurred if more time had been spent on defining the problem rather than hurrying to implement a solution. Most experts agree that the project is halfway complete once the real problem is defined, written down, and communicated.

Below is an example of the use of a Gantt chart for the development of a Web site for a small business.

Developing a Web Site

Jason and Melinda have a partnership to develop Web sites for small companies. A local martial arts school has asked them to develop a Web site to try to increase the school's business. Jason and Melinda meet with the owner to discuss the proposed Web site. During this meeting, they outline the following tasks, which will form the basis for the project:

- Determine the site requirements and needs of the school

- Select the name for the Web site and register the Web address

- Develop a tentative layout

- Develop content and obtain suitable graphics

- Contract with an Internet service provider to host the Web site on its servers

- Revise and fine-tune the Web site

- Ensure that the site goes live

- Follow up and arrange for periodic updates

- Review the plans with the customer during the development

After the meeting, Melinda develops the following Gantt chart for the martial arts Web-site project:

WEB-SITE DEVELOPMENT GANTT CHART

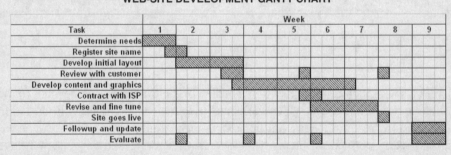

The Gantt chart graphically shows the progression of work required to complete the project.

Evaluate early.

Critical Path Management

Critical path management allows us to identify the critical points in the process. These critical points are readily identified by determining which tasks will cause substantial delays in the implementation of the solution if the schedule is not met. Suppose Jason and Melinda decide to use critical path management for the development of the Web site described in the preceding example. In the next figure, the bold lines and boxes indicate the critical path. It identifies items that require a fair amount of time to complete. If the schedule "slips," the Web site's "go live" date will be delayed. In this case, the development of the content and graphics with the customer takes the longest time and the site cannot be finalized until this task is completed; thus it is critical that this task be completed on schedule or the Web-site commissioning will be delayed. Noncritical path items, such as contracting with an Internet service provider (ISP), can be done as time permits after the critical items are completed.

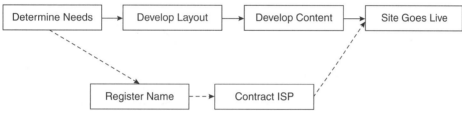

Critical Path Management of a Web-Site Development Project

A word of caution is in order here: If noncritical path items are completed too slowly, they can become critical path items. Thus using a critical path diagram is a dynamic process. The diagram should be continually updated as tasks are completed, so that you can view the overall progress of the project as a whole.

Coordination and Deployment

In most circumstances, groups of individuals will work together as a team to solve a problem. Under these conditions, coordination among various team members is imperative if the team is to achieve an efficient solution in the time allotted. The use of a **deployment chart** can help guide the team through the solution by assigning different team members either major or minor responsibilities related to each of the tasks. A deployment chart for the Web-site development project is shown here.

Deployment Chart for the Web-Site Development Project

Task	Team Member		
	Melinda	John	Web Programmer
Determine needs	▓	▓	
Register site name			▓
Develop initial layout		▓	
Review with customer	▓	▓	
Develop content and graphics			▓
Contract with ISP			▓
Revise and fine tune	▓	▓	
Site goes live			▓
Follow up and update	▓	▓	▓
Evaluate	▓	▓	

Proposed First-Year Budget for Web-Site Project

	Hours/Rate	Cost
Personnel		
Melinda, Project Director	40 hours @ $50/hour	$2000
Jason, Creative Designer	60 hours @ $50/hour	$3000
Web Programmer	60 hours @ $30/hour	$1800
Subtotal—Salaries		$6800
Monthly Maintenance/Updates		
Annual Fee	$100/month	$1200
Web-Site Name Registration		
Annual Fee	$25	$25
Internet Service Provider Web-Site Hosting		
Annual Fee	$25/month	$300
Supplies		
CDs for File Backup	$100	$100
TOTAL BUDGET		**$8425**

Necessary Resources

We must also estimate the resources necessary to complete the project. These resources usually fall into five categories: available personnel, equipment, travel, supplies, and overhead. Contingency funds are intended to cover unexpected expenses (e.g., permission fees for icons used on the Web site). At the beginning of a project, it is usually important to obtain an estimate of the total cost by preparing a budget similar to the one shown below. Often a project will not be approved for implementation before a budget is presented to decision makers.

The budgets for larger projects that are undertaken by larger companies may include fringe benefits—that is, charges for health and retirement programs for the personnel involved in the project. Sometimes "overhead" is charged on a project; overhead is used to defray organizational costs such as the salaries of the organization's management, building maintenance, and other general expenses.

Once the budget is submitted and approved, you enter the carry through phase of your problem's solution.

CARRY THROUGH

The **carry through phase** is an essential step in a successful problem-solving process. In this phase, the various people involved in the problem-solving process act upon the plans they have formulated. They may carry out a design, fabricate a product, conduct experiments, make calculations, prepare a report, and so forth. In some circumstances, the implementation process and the process of deciding the course of action are intertwined. For example, it might be necessary to collect experimental or other data (implement a plan) before the right decision can be made. Great care should be taken with the carry through phase, because all the planning in the world cannot compensate for a poor job of carrying through on the chosen solution. Here is a checklist of things to monitor in the carry through phase.

Carry Through Checklist

- ✓ Make an educated guess about what your solution will look like when you are finished. Fermi calculations are a great start.
- ✓ Make sure there is coordination of tasks and personnel.
- ✓ Constantly monitor your Gantt chart to make sure you stay on schedule.
- ✓ Evaluate each completed task along the way.
- ✓ Continue to learn as much as you can about the solution you have chosen.
- ✓ Read the literature and talk to your colleagues.
- ✓ Continue to challenge and/or validate the assumptions of the chosen solution. Make sure no physical laws are violated.
- ✓ Find the limits of your solution by creating simple models or making assumptions that would clearly both (1) overestimate the answer and (2) underestimate the answer.

LIST

✓ Construct a quick test or experiment to see whether the solution you have selected will work under the simplest conditions.

✓ Plan your computer experiments (i.e., simulations) as carefully as you would plan your experiments in the laboratory.

Flexibility is an essential trait for problem solvers to have if they are to deal with the inevitable changes that occur during projects.

FOLLOW UP

Inspect what you expect.

In the **follow-up phase,** we monitor our progress not only with respect to ensuring that the time deadlines are met but also with respect to ensuring that our solution does indeed solve the problem at hand. We periodically check the progress of the carry through phase to make sure it is meeting the following criteria:

- It follows the solution plan (that is, it meets the solution goals and fulfills the solution criteria).
- It is proceeding on schedule.
- It is staying within the budget.
- It is maintaining an acceptable quality.
- It is still relevant to solving the original problem.

It is important to monitor these points to confirm that the solution is "on track" and satisfies all the necessary goals. Be sure to check periodically that the problem is still correctly defined while you are implementing the solution because sometimes conditions or criteria can change during implementation that will invalidate the solution you originally chose.

Problem Statements That Change with Time

Sometimes it may feel as if you are shooting at a moving target, as the desired goals change over the course of the project. The problem statement may change for a number of reasons—for example, because of changing market conditions. The introduction of a competing product or service, reduced financing, or other factors could affect either problem statements or planned solutions. If during carry through, evaluation, or any of the other phases some or the entire project cannot be accomplished, the problem statement or solution must be modified. Of course, this type of information is learned only *after* we begin developing the solution. An example of adapting during the problem-solving process is detailed in the following.

Adapting during the Problem-Solving Process: Better TV

Bracken Technologies was in the process of designing a new 40-inch TV that it was hoping would increase the company's profitability. It would have market leading picture quality for its size and be sold for $1200. During the course of the company's product's development, one of the engineers did follow-up research on the TV market. He found out about a competitor launching a 40-inch TV model with even better picture quality for nearly the same price.

The engineer reported his finding to the project team and they returned to the idea generation step in the problem-solving process. They came up with several ideas and were then faced with several alternatives, which included cutting the TV's price to significantly undercut the price of the competitor's new product, or improving the TV's picture quality to surpass the design of the competitor's new product. The company eventually was able to figure out a way to use the picture quality technology in a larger TV for a similar price and launched a 60-inch TV with the same picture quality to become the market leader in the more lucrative 60-inch TV segment of the market.

GOAL SETTING

Throughout the entire process of implementing the solution quality, goal setting is vital. Goals should pass the SMAC test. This means that goals should be *specific, measureable, attainable,* and *compatible.* For example a college student entering her final year is looking into possible job opportunities. She sees that many companies require a 3.0 GPA, with some others as high as 3.5. The student currently has a 2.9 GPA and decides to set a goal of raising her GPA to 3.1 by the end of the year. This goal is *specific*: she has a number set rather than saying she wants to do well in school. This goal is *measurable*: she will be graded in her classes and get the results. This goal is *attainable*: raising her GPA 0.2 in one year can be done, whereas raising it 0.6 for the other jobs would not be possible due to the limited effect one year has on her overall GPA. Finally, this goal is *compatible* with becoming eligible for job opportunities with a 3.0 minimum GPA and her 3.1 would qualify her for those, whereas maintaining her GPA would not.

SUMMARY

Many people get stalled in the problem-solving process because they get bogged down in the planning process and forget the most critical step—implementation. In this chapter, we presented a number of techniques to facilitate the implementation process.

- **Approval**

  - Sell your ideas for the project.
  - Make the case for change.

- **Planning**

  - Sketch the pathway through to the solution.
  - Plan for appropriate resource allocation.
  - Perform a K.T. situation appraisal and potential problem analysis.
  - Prepare Gantt and deployment charts.
  - Identify the critical tasks and prepare a critical path diagram.
  - Prepare a budget for your project.

- **Carry Through**

  - Monitor the progress of the critical tasks closely.
  - Use the carry through checklist.

- **Follow Up**

  - Confirm that the solution meets the specified objectives and criteria.

WEB-SITE MATERIAL (WWW.UMICH.EDU/~SCPS)

- **Learning Resources**
  Summary Notes

- **Interactive Computer Modules**

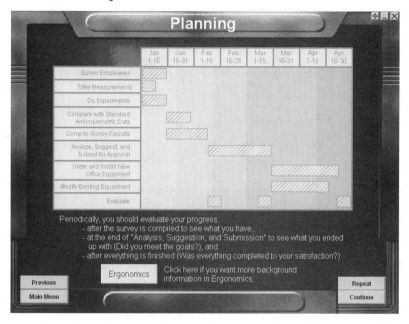

- **Professional Reference Shelf**
  1. Thanksgiving Dinner Example
  2. The Minimum Number of Experiments (or . . . "Getting the Most Bang for Your Buck")
  3. Popcorn Example
  4. Heat Exchanger Example
  5. Revealing the Solution
  6. Critical Path Management
  7. What Happens When the Goal Keeps Changing?
  8. Experimental Projects

LIVERPOOL JOHN MOORES UNIVERSITY
LEARNING SERVICES

REFERENCES

1. Peters, M. S., K. D. Timmerhaus, and R. E. West, *Plant Design and Economics for Chemical Engineers*, 5th ed., McGraw-Hill, New York, 2002.
2. Strunk, W., and E. B. White, *The Elements of Style*, 4th ed., Macmillan, New York, 1999.

EXERCISES

9.1. List the five most important things you learned from this chapter.

9.2. Prepare a carry through checklist for the situations in the following examples:

A. "25 Years or Less?" (Chapter 9)

B. "Changing Buses" (Chapter 9)

C. "Choosing a Paint Gun" (Chapter 8)

D. "Decision Analysis: Popara Enterprises Picks a New Market" (Chapter 8)

9.3. Prepare a follow-up checklist for each of these situations:

A. "K.T. Potential Problem Analysis: Jet-Lagged John" (Chapter 8)

B. The example regarding the dead fish in the river (Chapter 3)

9.4. Choose one of the situations in the following table and carry out one or more of the following activities:

A. Prepare a Gantt chart for the activity.

B. Prepare a deployment chart for the activity.

C. Prepare a critical path-planning chart for the activity.

D. Use Bloom's taxonomy (see the Professional Reference Shelf on the Web site) to outline how you would carry through on the activity.

E. Prepare a budget for the activity.

Situations for Exercise 9.4

• Planning a surprise birthday party	• Getting admitted to medical school
• Planning a wedding	• Obtaining an overseas internship or co-op
• Planning a camping trip to Colorado	
• Getting elected mayor of your city	• Becoming a commercial airline pilot
• Publishing your autobiography	• Getting cast on a reality television series
• Becoming a U.S. Supreme Court justice	• Selecting and financing a new car
• Building a new home	• Deciding on and arranging a cruise in a foreign country

9.5. Prepare a Gantt chart, a deployment chart, a critical path-planning chart, and a budget for a different activity of your choice.

9.6. Prepare a write-up (or presentation) to management for the following proposals:

   A. You want to attend a professional meeting or take a short course in Europe.

   B. You want your company to market a new widget.

9.7. You are going to prepare a three-course dinner for your gourmet dinner group for a party of eight.

Course	Item	Preparation Time	Eating Time
Appetizer	Bacon-wrapped water chestnuts	Cook in oven 10 minutes	10 minutes
Soup	Onion soup	Cook 30 minutes on stove	15 minutes
	Bread sticks	Warm 10 minutes in oven	
Entrée	French pot roast	Cook 2 hours in oven	40 minutes
	Mashed potatoes	Cook ready mix 10 minutes	
	Fresh mixed vegetables	Boil 20 minutes	
	Gravy	Cook juice from roast 10 minutes on stove	
Dessert	Apple pie	Cook 35 minutes in oven	15 minutes
	Ice cream	Let stand 5 minutes before scooping	

Prepare a Gantt chart, critical path flow chart and a stove deployment chart (i.e., oven top) for your dinner party.

9.8. You just won the State Lottery for $5.3 M. Decide what to do, and prepare an implementation plan.

FURTHER READING

Blanchard, Kenneth, and Robert Lorber. *Putting the One Minute Manager to Work: How to Turn the Three Secrets into Skills.* Berkeley Books, Berkeley, CA, 1984. Increase your productivity using three easy-to-follow techniques.

Hendrix, Charles D. "What Every Technologist Should Know about Experimental Design." *Chemtech*, p. 167, March 1979. Some useful background material on designing experiments to gain the maximum amount of information with the least amount of effort.

Massinall, John L. "The Joys of Excellence." *Chemtech*, 20, p. 393, July 1990. Timely tips for managing a project and keeping work on schedule.

McCluskey, R. J., and S. L. Harris. "The Coffee Pot Experiment: A Better Cup of Coffee via Factorial Design." *Chemical Engineering Education,* p. 151, Summer 1989. More interesting background on designing experiments.

Murphy, Thomas D. "Design and Analysis of Industrial Experiments." *Chemical Engineering*, 84, p. 168, June 6, 1977. Excellent overview of factorial design of experiments.

Starfield, Anthony M., Karl A. Smith, and Andrew L. Bleloch. *How to Model It: Problem-Solving for the Computer Age.* McGraw-Hill, New York, 1990. Chapter 6 of this book contains a nice description and additional examples of critical path planning.

10   EVALUATION

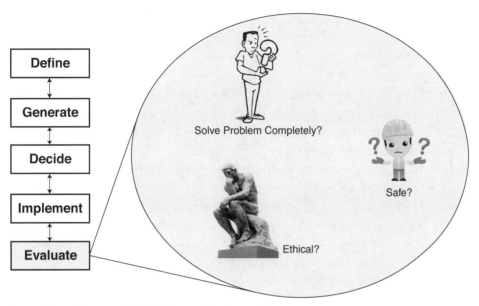

Solve Problem Completely?

Ethical?

Safe?

(Images from Shutterstock: Rob Wilson and andartenot)

Evaluation of our solution to the problem is the last task in our problem-solving heuristic. In this chapter we present guidelines for evaluating a solution to make sure it completely solves the problem specified, is ethical, and is safe for both people and the environment.

GENERAL GUIDELINES

Evaluation should be an *ongoing* process throughout the life of a project. As each phase of the project is completed, we should examine the goals and accomplishments of that phase to make sure they were satisfied before we proceed to the next phase. We should also evaluate future directions in light of the results of each phase to verify that the direction in which we are proceeding is still the correct one. We must look for any fallacies in logic that might have occurred, especially at key decision points during the project. In addition, we must challenge the various assumptions that were made. Have all unstated assumptions been recognized? For example, in our earlier case study in Chapter 3, was the engineer justified in assuming that the managers of the plant no longer have to worry about the effluent waste stream no matter how low the river water level becomes because the fish were dying from a fungus?

Evaluation Checklist

To address these types of questions, have someone outside the group that developed the solution review the assumptions and solution logic. They have a fresh perspective and can often catch things that you might overlook. During the evaluation process, you must inevitably make qualitative and quantitative judgments about the extent to which the material and methods satisfy the external and internal criteria. Ask evaluation questions such as those given in the following checklist.

Evaluation Checklist

LIST

- ✓ Is the solution logical?
- ✓ Does the proposed solution solve the real problem?
- ✓ Is the problem permanently solved, or is this a patchwork solution?
- ✓ Are all the criteria and constraints satisfied?
- ✓ Does the solution have the desired impact?
- ✓ Is the solution economically, environmentally, politically, and ethically responsible and safe?
- ✓ Have you used the different types of Socratic questions—for example, have you challenged the information and assumptions provided?
- ✓ Have all the consequences of the solution been examined—for example, does it cause other, more serious problems?
- ✓ Have you argued both sides—the positive and the negative?
- ✓ Has the solution accomplished all it could?
- ✓ Is the solution blunder-free?
- ✓ Have you checked the procedure and logic of the arguments?

You need to confirm *all* findings. Check whether any piece of the puzzle (i.e., the solution) doesn't fit and consequently may require the entire solution to be greatly modified or perhaps even scrapped.

Evaluation of Bird Droppings and Car Paint

A new Australian auto dealership recently began experiencing increasing damage to the cars kept on its lot. Previously, the cars spent very little time out in the open lot as they were sold almost as soon as they arrived. However, recent sales had slowed, and the cars spent a longer time on the outdoor lot waiting to be sold. While out on the open lot they were being hit by bird droppings. The car polish used at the time did not remove the stains completely, and the stains were still visible after washing and buffing, thereby further reducing sales and increasing maintenance costs. To tackle this problem, the manager assigned the following task to his technical specialist: *Find a better car polish that can remove the stains caused by bird droppings.* After collecting responses and observations, it was found that this problem seemed to occur more often on hot sunny days. The specialist questioned whether this occurrence on hot days was related to the true cause of the problem.

To follow up on the previous observation, that stains occur more often on hot sunny days, the specialist decided to go online and search for the composition of bird droppings and car paint to investigate how they affect one another. Upon initial research, he learned that birds excrete uric acid in their droppings. One of his colleagues remembered that acids are very corrosive and called an old friend, who majored in material science, to verify this fact. The material science friend told him that it is difficult to develop a car paint to completely prevent the bird droppings from damaging the car, and removing the stains is even tougher.

After analyzing the collected information, they decided that the real problem may not be related to the kind of car polish used, but instead they had to prevent car paint damage by preventing contact between bird droppings and the car paint altogether.

After consulting and brainstorming with several workers and management, three proposed possible solutions for the problem were examined: expand indoor storage, install tarps individually, and keep a few falcons at the dealership to scare away the masses of birds.

The management at the dealership carried out a K.T. decision analysis after brainstorming the three possible solutions. The musts were that the solution must be low in cost and effective, and the wants were easy to implement, long-term durability, resistance from weather, and minimal interference with car sales. The car dealer wants a solution that is easily implemented. Both the tarps and falcons solutions are easy to implement. The falcons are a bit easier, because the solution does not involve doing something to every single car.

So the falcons receive a 9 for ease of implementation, while the tarps get an 8. The tarps and falcons score the same high rating (9) for long-term durability. For weather resistance, the tarps receive a 10, while the falcons are much more susceptible to the effects of bad weather and receive a rating of 6. Tarps must be removed to show the vehicles to prospective customers and do not present a very attractive looking sales lot. The falcons solution has none of those drawbacks and thus the falcons receive a high rating of 9 for this want, while the tarps receive a much lower rating of 3.

Options / Musts		Expand indoor storage.		Install tarps on every car.		Keep falcons to scare off birds.	
Effective		Yes		Yes		Yes	
Low in cost		No		Yes		Yes	
Wants	**Weight**	**Rating**	**Score**	**Rating**	**Score**	**Rating**	**Score**
Easy to implement	6			8	48	9	54
Long-term durability	8			9	72	9	72
Resistance from weather	5			10	50	6	30
Minimal interference with car sales	10			3	30	9	90
Total		-			200		246

After arriving at the initial conclusion to have falcons at the dealership, the risks involved in each option were evaluated by using an adverse consequences analysis and it was determined that the falcons were still the optimal choice.

Alternative	Probability (P)	Severity (S)	Threat (P × S)
Tarps			
Less appealing	10	7	70
Employees spend more time dealing with tarps	6	8	48
Total			**118**
Keeping falcons			
Falcons may not fly over the outer part of the lot	2	10	20
Falcons' droppings	2	10	20
Total			**40**

The K.T. decision analysis demonstrated that the best method of preventing stains from bird droppings was to purchase falcons and keep the birds away from the outdoor cars. No birds, no droppings! Although they would still have to deal with a few falcon droppings, the reduction in total droppings would be significant due to the reduction from hundreds of birds to one group of falcons. The dealership then used the evaluation checklist to make sure the solution completely solved the real problem.

✓ *Were the information and assumptions provided challenged?* The assumption that the polish was the problem was challenged, and that led to finding the real problem of keeping the droppings off the cars altogether. The assumption that the bird droppings were responsible for the damage was challenged and then confirmed. It was also known that falcons could effectively keep birds away from the cars, since they are predators.

✓ *Did the solution solve the real problem? What was the real problem in this case?* The real problem was not inefficient car polish, as originally thought, but rather that during hot sunny days the presence of acidic bird droppings caused severe damage to the cars' coating paint. The falcons scared the birds away and eliminated the bird droppings on the cars.

✓ *Was the problem permanently solved?* By scaring away the birds, droppings stopped landing on the car paint, solving the problem permanently as long as the falcons were around. The falcons are living wild animals, so what is stopping the falcons from flying away? This question caused the employees at the dealership to learn how to train the falcons to stay living at the dealership.

✓ *Did the solution have the desired impact?* Yes, the solution prevented bird droppings from damaging the cars and was observed immediately after purchasing the falcons.

✓ *Were all consequences of the solution examined?* Yes, the solution using falcons was designed to anticipate many adverse consequences. The dealership realized it should let customers know about the falcons so the customers were not scared away. The manager put up a sign talking about the falcons and the customers seemed to enjoy trying to spot them around the dealership lot.

✓ *Are all criteria and constraints satisfied?* The criterion of effectively preventing damage for an extended time was satisfied. In addition, constraints such as the size of the lot and budget were satisfied as well.

Based on a true example.

ETHICAL EVALUATION

We are now going discuss three models you can use to help you see if there are any ethical issues you need to consider in your solution: Four Classical Virtues, Ethics Checklist, and the Five P's.

The Four Classical Virtues

Seebauer and Barry[1] provide a simple model for the origin of moral action, in which *emotions and mind* feed into *will, decisions, and actions*. They also discuss how these components depend on four classical virtues:

- **Prudence:** Thinking about a moral problem clearly and completely
- **Temperance:** Avoiding either being rash or suppressing our emotions
- **Fortitude:** Not moving blindly away from something we do not like
- **Justice:** Having the will to act in truth on the way things actually are and to act with fairness to all concerned

After discussing the four virtues in detail, these authors put forth a key principle: "People should always decide and act according to these virtues as far as possible." Thus Seebauer and Barry describe another way of looking at the ethical decision-making process—namely, as a "four-component model":

- Sensing the presence of moral issues (I had not thought about . . .)
- Reasoning through the moral issues (the four virtues and the five P's)
- Making a decision
- Following through on the decision

We could also use the K.T. decision analysis technique in Chapter 8 to help us make the decision and the implementation procedure discussed in Chapter 9 to help us carry through on our solution.

Ethics Checklist

In their book *The Power of Ethical Management*,[2] Blanchard and Peale offer a set of guidelines to help us quickly sort through the issues at hand and reach an ethical solution. These guidelines also help us to uncover the moral issues that the decision entails by providing the five P's and a checklist of questions to consider.

Ethics Checklist[2]

LIST

- ✓ Is it legal? Will I be violating either civil law or company policy?
- ✓ Is it balanced? Is it fair to all concerned in both the short term and the long term? Does it promote win/win relationships?
- ✓ How will it make me feel about myself? Will it make me proud? Would I feel good if my decision were published in the newspaper? Would I feel good if my family knew about it?

"There is no pillow as soft as a clear conscience."
—John Wooden, UCLA Bruins

If the answer to the first question could be interpreted from any viewpoint or appearance as "No, it is not legal," then there is no need to proceed to the second and third questions. However, if the solution is legal and does not violate company policy, then the second question raises the flag if a decision greatly benefits one person or company, but unfairly takes advantage of others. That is, it may eventually come back to haunt that individual or company (e.g., excessive interest rates on overdue credit cards). Blanchard and Peale's last question is meant to activate our sense of fairness and make sure that our self-esteem is not eroded through an unethical decision. This ethics checklist helps us address one of the knottiest problems in business: *How can we get acceptable bottom-line results and stay competitive, while at the same time making sure we are being ethical?*

The Five P's

Blanchard and Peale identify five P's that need to be considered in analyzing the solution to the problem at hand: purpose, pride, patience, persistence, and perspective. The five P's table gives a list of questions for us to answer that will help us further evaluate our solution.

To facilitate the ethical evaluation process, try asking someone to critique your answers to each of the questions for the five P's. This person (called an advisor) could play a passive role—simply listening to your explanation—or an aggressive role—actively questioning your every point. Even when the advisor listens in a passive mode, the mere fact that you must verbalize the application of the five P's to your situation will help improve your evaluation process.

The Five P's[2]

Purpose:	What is the objective for which you are striving? Are you comfortable with that as your purpose? Does your purpose hold up when you look at yourself in the mirror? (This P involves the virtue "prudence.")
Pride:	Can you take pride in the solution you have developed? Is there any false pride or self-doubt involved? (This P can be related to the virtue "justice.")
Patience:	Have you taken the time to think through all the ramifications of your solution? (The virtue "temperance" plays a role in this P.)
Persistence:	Are you sticking to your guns and not being dissuaded by other demands? Have you given up too soon on finding a solution that is fair and balanced to all concerned? (This P draws from the virtues "temperance" and "fortitude.")
Perspective:	Have you taken the time to focus inside yourself to be sure everything fits with your ideals and beliefs? How does the solution fit into the "big picture"? (This P can be related to the virtues "prudence" and "justice.")

Perspective (the fifth P) is the hub around which the other P's rotate. Part of perspective is developing the inner guidance that is awakened by the other P's and that helps us see things more clearly.

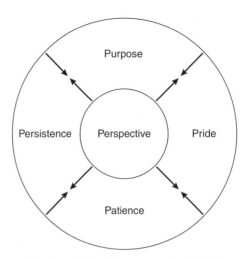

The Interrelationships among the Five P's

"The greatest battles of life are fought out daily in the silent chambers of the soul."
—David McKay

Case Study Examples

As discussed earlier, we should carry out an evaluation at each of the key decision points and not wait until the very end of the project to learn that perhaps the solution path that we chose was unethical. Let's consider some examples to see how this process works.

The Holiday Gift[3]

Henry is in a position to influence the selection of suppliers for the large volume of equipment that his firm purchases each year. At holiday time, he usually receives small tokens from several salespeople, ranging from inexpensive ballpoint pens to a bottle of liquor. This year, however, one salesperson sent an expensive briefcase stamped with Henry's initials. This gift is very much out of the ordinary.

What should Henry do?

1. Keep the case, on the grounds that his judgment will not be affected in any way?
2. Keep the case, because it would simply cause embarrassment all around if the case were returned?
3. Return the case?
4. Other? (Please specify.)

The Five P's

Purpose: Ask yourself what you would do if you were in Henry's shoes to remain unbiased in selecting the best supplier for a given job. How would your accepting the gift be preceived by others inside and outside you company? Could you make a list of the pros and cons of accepting the gift?

Pride: Would you feel pride in accepting the case or pride in returning the case? How would it make the salesperson feel if you returned it? Will he feel that he is dealing with a very ethical company? Or would it make him feel sheepish that he had done something improper?

Patience: Set aside a time to think about whether you should accept the case. Talk to someone whose judgment you trust.

Persistence: Have you pursued all avenues to resolve the issue of either keeping or returning the case? Have you investigated the company's guidelines (e.g., dollar value) of accepting gifts?

Perspective: Even if you feel your judgment will not be affected by accepting the case, how will it appear to other colleagues? Are you setting a good example?

The following is the response to a reader survey carried out by *Chemical Engineering*. The majority of the people (65.8%) thought the case should be returned. However, 27.7% of those respondents who were younger than age 26 thought it would be okay to keep the case.

elisekurenbina/
Shutterstock

Responses by Age (%)				
All Ages		**<26**	**26–50**	**>50**
Option 1	19.7	27.7	17.9	15.4
Option 2	3.3	4.9	2.8	3.1
Option 3	65.8	56.9	68.1	67.6
Option 4	9.5	9	9.8	12.1

Comments by Respondents to the Reader Survey

- "Keep the case and really not be affected by it, which means that the next holiday you're back on ballpoint pens."

- "As a procurement agent on a limited scale, I have heard of colleagues receiving Porsches. The gifts only get bigger if you accept the first. Eventually, it will affect your judgment."

- "I sometimes feel guilty accepting a gift knowing that the giver won't get anything out of it. But I won't be bribed."

- "My price is very much higher than a briefcase. As a matter of fact it is so high that it would not be profitable to meet it."

- "Henry risks being fired for a crummy $80 briefcase. That's the trouble—it's never two fully paid round-trip tickets to Hawaii, it's always cheap junk."

Other Thoughts

The above bullet points are all good ones. Henry's best option and path forward may be to research the company policy on gifts and if it is over the allowed limit discuss it with his supervisor on how to best handle the situation. If the gift is returned, it could be with a note saying he cannot violate the company rules.

Rigging the Bidding[3]

Steve is a design engineer for a large chemical company. There are two competitive pieces of equipment that are sold to do a job required by the process he is working on. Equipment from ABC Inc. is widely advertised and sold, but Steve has heard through the grapevine from his competitor's plant that ABC's equipment tends to break down unexpectedly and often. However, there is no way for him to document this information. XYZ Corporation makes equipment that will do the same job but is much more expensive. Steve knows from his own experience that this equipment is quite reliable. It is company policy to obtain competitive bids, and in such a situation ABC would certainly win. Steve deliberately rigs the specifications by inserting unnecessary qualifications that only the XYZ equipment will meet.

1. Is Steve being ethical in using false specifications to circumvent company policy, even if he believes it is in the company's best interests?
2. Would you do the same if you were in Steve's place?

The Five P's

Purpose: To choose the best piece of equipment for your company.

Pride: Will you feel pride in having reached the best decision and the way in which you reached it?

Patience: Have you taken the time to check with people in the company to seek their advice about which way to go?

Persistence: Have you checked the grapevine to see if it is a rumor or a fact that the equipment is unreliable?

Perspective: Are you setting a precedent so that you might write future specifications for only one company?

Responses from the reader survey in *Chemical Engineering*:

Question 1: Responses by Age (%)				
All Ages		<26	26–50	>50
Yes	36.8	40.8	17.9	32.8
No	55.7	53.5	57	54.9

Question 2: Responses by Age (%)				
All Ages		<26	26–50	>50
Yes	30.5	35.9	30.1	24.4
No	61.2	56.9	62.6	61.5

Comments by Respondents

- "Can a professional judgment be made based on the grapevine?"
- "Go get that grapevine story verified or busted. Get the facts."
- "Anecdotal reports are a way to document suspicions. One must be careful that one isn't depending on a few disgruntled users; but, surely, it's reason to do more thorough research."
- "To eliminate red tape within a large company and obtain the best equipment, one must often change the rules."
- "Since it'd be me that was called at 2 a.m. when the ABC equipment broke, I'd probably force the choice of XYZ."
- "I have not heard of any design engineer being made to decide based on price and specifications alone. Judgment is a highly sought quality in an engineer. Sometimes, gut feeling counts, too."

SAFETY CONSIDERATIONS

One of the most important parts of the evaluation process is to make sure the proposed solution is safe for both humans and the environment. Carrying out a potential problem analysis (see Chapter 8), a fault-tree analysis (see the kerosene heater problem on the Professional Reference Shelf on the Web site), or a hazardous operations analysis (HAZOP) is helpful to make sure you have considered all aspects of the solution that might affect safety. See http://www.sache.org/ for more information on process safety.

SUMMARY

Evaluation of the proposed solution to the problem is the last task in the problem-solving heuristic. Evaluation should be an ongoing process through-out the life of a project. Use the evaluation checklist to make sure you have done the following:

- Challenged information and assumptions
- Solved the *real* problem
- Examined all consequences of the solution
- Found a solution that is logical, ethical, safe, and environmentally responsible

Seebauer and Barry provide a simple model for the origin of moral action, in which emotions and mind feed into will, decisions, and actions. Their four-component model involves the following activities:

- Sensing the presence of moral issues
- Reasoning through the moral issues
- Making a decision
- Following through on the decision

These components depend on four classical virtues: prudence, temper-ance, fortitude, and justice.

You can use the ethics checklist and the five P's to make sure your decision and solution are ethical. The five P's are purpose, pride, patience, persistence, and perspective.

WEB-SITE MATERIAL (WWW.UMICH.EDU/~SCPS)

- **Learning Resources**
  Summary Notes

- **Professional Reference Shelf**
  Ethics: Different Versions of Buffalo Nation
  Additional Ethics Problems
  Different Versions of Buffalo Nation

REFERENCES

1. Seebauer, Edmund G., and Robert L. Barry, *Fundamentals of Ethics for Scientists and Engineers,* Oxford University Press, New York, 2007.
2. Blanchard, Kenneth, and Norman Vincent Peale, *The Power of Ethical Management*, Ballantine Books, Fawcett Crest, New York, copyright ©1988 by Blanchard Family Partnership and Norman Vincent Peale.
3. *Chemical Engineering*, p. 132, September 1980.

EXERCISES

10.1. Prepare an evaluation checklist that would help determine if the "25 Years or Less?" initiative discussed in Chapter 9 was working.

10.2. A. "Should You Kiss a Co-worker?" Carry out an ethics analysis on the following example after viewing "When should you kiss your co-worker?" at http://mconnex .engin.umich.edu/brownbaglearning/2012/when-you-should-kiss-your-co-worker-leading-across-cultural-divides/.

　　　B. Please analyze the situation below from all sides and write a one- to two-page analysis. State your position and defend it.

　　　Your company has been working hard to capture business in a non-western country. Your representatives on the scene have made a major breakthrough and it looks like a plant will be built as a joint venture. It would be a close copy of a facility just completed in the United States. The project manager for the U.S. project is available and is clearly the choice to head the delegation to the foreign site. The problem is, the project manager is female and the culture in the foreign country in question believes that females should be restricted to traditional roles. The project manager is clearly the most qualified individual, both in advancing the project *and* in ensuring that your company's interests are being advanced. But there is a serious question of how her inputs will be received at the foreign site.

　　　1. You are the VP of Technology. You must assign the team to go to the foreign country. Discuss the parameters involved in your decision from all sides. Define YOUR position and defend it.

2. You are the project manager. Discuss your options. Define YOUR position and defend it.

Note: Ethics are personal interpretations of the proper course of professional conduct in a given situation. While there are "norms" for ethical decisions, they are very much influenced by individual perceptions and therefore can justifiably show significant variations.

You will be graded on your analysis and defense of your ethical positions on the above issue, not on the positions themselves.

10.3. Nick is chief engineer in a phosphate fertilizer plant that generates more than one million tons per year of gypsum, a waste collected in a nearby pile. Over many years, the pile has grown into a mountain containing 40 million tons of waste. There is little room at the present site for any more waste, so a new gypsum pile is planned.

Current environmental regulations call for the elimination of acidic water seepage and groundwater contamination by phosphates and fluorides. Nick's design for the new pile, which has been approved, incorporates the latest technology and complies with U.S. Environmental Protection Agency and state regulations. However, he also knows that the old pile—although exempt from the current regulations—presents a major public hazard. When it rains, acidic water seeps through the pile, carrying phosphates into the groundwater.

In a confidential report to management, Nick recommends measures that will prevent the seepage from happening. His company turns down his proposal, stating that, at present, no law or regulation demands such remedy. Use the four virtues, four-component model, evaluation checklist, ethics checklist, and five P's to help you analyze the situation.

Problem adapted from *Chemical Engineering*, p. 40, March 2, 1987.

10.4. The environmental and safety control group in a circuit-board etching and plating plant has just completed a program to improve the measurement of toxic releases into the atmosphere in response to stricter regulations recently issued by the state health and environmental commission.

Small amounts of a toxic material are detected for the first time by means of a new instrument purchased and installed at the suggestion of Joan, the group leader. The detection method specified by the state does not reveal any trace of chemical.

A search through books and magazines shows that this material is not dangerous in the low concentrations detected, although the state agency says it is, basing its claim on the extrapolation of published data. Use one or more of the following heuristics—four-component model, evaluation checklist, ethics checklist, and/or five P's—to help you address this situation.

Problem adapted from *Chemical Engineering*, p. 40, March 2, 1987.

10.5. Each week, operators in a plant regularly dump more than 100 drums of a dry, dusty enzyme (isomerase, used in converting dextrose to fructose) into a large tank, where it is rehydrated for several hours. The dust, although not immediately harmful, is suspected of causing long-term allergies and even lung problems. However, these symptoms show up in fewer than 10% of people who are exposed to the enzyme.

The plant safety code requires that operators wear masks, goggles, and gloves when dumping the enzyme. Because the temperature in the working area is usually about 110°F, the operators often ignore this requirement.

This is the situation found by Phil, who has very recently taken over supervision of the department. Use one or more of the following heuristics—the evaluation checklist, ethics checklist, and/or five P's—to help you address this situation.

Problem adapted from *Chemical Engineering*, p. 40, March 2, 1987.

10.6. You are employed by a small company that is trying to build a plant to produce chemical X. As a result of its low overhead and other factors, your firm should be able to significantly undercut the price charged by your competitor for the same chemical. You have sized all the pieces of equipment except the reactor. You cannot do this because you don't know the kinetic parameters, nor do you have laboratory equipment to determine them. Time is of the essence, so your boss suggests that you take a photograph of your competitor's reactor, which is located outside its plant, and use that information to size your reactor. He suggests you hire a plane to take aerial photographs or a truck similar to those used to repair telephone lines that could see over the fences. Your boss also suggests you could get an estimate of the production rate by monitoring the size and number of trucks shipping the product from your competitor's plant.

How confident are you that your estimates of the sizes will be reasonably accurate? Do you believe it is ethical to estimate the reactor sizes and production rates in this manner? Make a list of the reasons and arguments as to why your boss might think it is ethical to make this request. If you don't feel it is ethical, make a list of reasons and arguments as to why the kind of surveillance suggested by your boss should not be done. Suggest alternative ways to obtain the desired information. You may wish to consult the books *Fundamentals of Ethics* by Seebauer and Barry and *The Power of Ethical Management* by Blanchard and Peale to identify and evaluate ethical issues.

If you believe that the situation described here is clearly ethical or unethical, revise the scenario so that it falls into a gray area. For example, would obtaining the needed information in this way be ethical if the reactor were in full view from the street? What if your boss suggested that you tour the plant with a Boy Scout troop and try to take pictures and obtain other information (e.g., read gauges) while on the tour at the competitor's annual "Engineering as a Profession Day" three weeks from now? Use one or more of the following heuristics—the four virtues, four-component model, evaluation checklist, and/or ethics checklist—to help take the "grayness" out of the situation and make it black or white.

10.7. Jay's boss is an acknowledged expert in the field of data analysis. Jay is the leader of a group that has been charged with developing a new catalyst system. So far the group has narrowed the candidates to two possibilities: catalyst A and catalyst B.

The boss is certain that the best choice is A, but he directs that the tests be run on both catalysts "just for the record." Owing to the fact that inexperienced employees run the tests, the tests take longer than expected, and the results show that B is the preferred material. The engineers question the validity of the tests, but because of the project's timetable, there is no time to repeat the series. The boss directs Jay to work the math backward and come up with phony data to justify the choice of catalyst A—a choice that all the engineers in the group, including Jay, fully agree with. Jay writes the report.

What would you do?

A. Write the report as directed by the boss?

B. Refuse to write the report, because to do so would be unethical?

C. Write the report, but also write a memo to the boss stating that what is being done is unethical—to cover you in case you are found out?

D. Write the report as directed, but refuse to have your name on it as the author?

E. Go over your boss's head and report that you have been asked to falsify records?

F. Do something else? (If so, what?)

Use one or more of the following heuristics—the four virtues, four-component model, evaluation checklist, ethics checklist, and/or the five P's—to help you analyze each of these options.

Problem adapted from *Chemical Engineering*, p. 132, September 1980.

10.8. In Exercise 10.7, Jay decided to write the report to suit his boss, and the company went ahead with an ambitious commercialization program for catalyst A. Jay has been put in charge of the pilot plant where development work is being done on the project. To allay his doubts, he personally runs some clandestine tests on the two catalysts. To his astonishment and dismay, the tests determine that while catalyst A works better under most conditions (as everyone expected), at the operating conditions specified in the firm's process design, catalyst B is, indeed, considerably superior.

If you were Jay, what would you do?

A. Since no one knows that you've done the tests, keep quiet about the results because the process will run acceptably with catalyst A, although not nearly as well as it would run with catalyst B?

B. Tell your former boss (the catalyst expert) about the clandestine tests and let him decide what to do next?

C. Make a clean breast of the whole affair to upper management, knowing that it could get you and a number of colleagues fired or, at least, discredited professionally?

D. Do something else? (If so, what?)

Use one or more of the following heuristics—the four virtues, four-component model, evaluation checklist, ethics checklist, and/or the five P's—to help you analyze each of these options.

Problem adapted from *Chemical Engineering*, p. 132, September 1980.

10.9. Ruth works as a group leader for a company that sells large quantities of a major food product that is processed before sale, by heating. Her product-development group has been analyzing the naturally occurring flavor constituents of the product, and Ruth has discovered that several of the flavor components (actually pyrolysis products, present in minute quantities) are chemicals that have been found to cause cancer in animals when given in large doses. Yet, the product—in worldwide use for literally centuries—has never been implicated as a cause of cancer.

Although in the United States the Delaney Amendment to the Pure Food and Drug Act prohibits companies from adding cancer-causing agents to food, there are no government regulations concerning those that may occur naturally.

What should Ruth do?

A. Quash the report?

B. Submit a confidential report to her superiors and let them decide?

C. Submit an article summarizing the compounds found to a reputable journal, but without mentioning cancer?

D. Notify a consumer-protection organization?

Use one or more of the following heuristics—the four virtues, four-component model, evaluation checklist, ethics checklist, and/or the five P's—to help you analyze each of these options.

Problem adapted from *Chemical Engineering*, p. 132, September 1980.

10.10. The company that employs Reginald has a practice of using salaried personnel to replace striking workers and paying these people double-time pay for any work in excess of 40 hours per week, plus a $100-per-day strike bonus. (Under ordinary circumstances, overtime pay is never granted to salaried personnel, which includes engineers.) Not having a union themselves, Reginald and his fellow engineers have been hard hit by inflation, and many would welcome the opportunity to earn extra pay.

The plant is presently being struck by union operators over "unsafe" working conditions, which Reginald personally believes may be unsafe but which are not covered specifically under government safety regulations. The company disputes the union's contention about safety. The strike looks as if it could be a lengthy one.

What should Reginald do?

A. Refuse to work, because he thinks the union's allegations may have merit?

B. Refuse to work, because he believes that strike breaking is unethical?

C. Work, because he believes that filling in for striking workers is an obligation of all members of management?

D. Work, because the extra pay is a great way to catch up on some of his bills or earn the down payment on a car?

E. Work, because he believes he may be fired if he doesn't?

F. Do something else? (If so, what?)

Use one or more of the following heuristics—the four virtues, four-component model, evaluation checklist, ethics checklist, and/or the five P's—to help you analyze each of these options.

Problem adapted from *Chemical Engineering*, p. 132, September 1980.

10.11. Larry's company has been using a flavor additive in one of its products, but there have been problems with the flavor's stability. One of Larry's chemists accidentally discovers that the flavor can be stabilized by adding a mixture of tin and lead in very small quantities. Although both tin and lead are recognized poisons, the chemist points out the amounts added are not more than the amounts that might be leached out of the soldered seams of the common tin cans used for a multitude of food products. The new product will be packed in glass, so no further addition of heavy metals will occur.

What should Larry do?

A. Recommend that the additive not be used, because it is unethical to add poisons, no matter what the quantity?

B. Prevent any further problems by suppressing the finding?

C. Recommend the open use of this heavy-metals-stabilized additive?

D. Recommend that the tin and lead additives be used, but that the deliberate addition of heavy metals be considered a trade secret and kept from leaking to the public because "it would only cause unnecessary worry"?

Use one or more of the following heuristics—the four virtues, four-component model, evaluation checklist, ethics checklist, and/or the five P's—to help you analyze each of these options.

Problem adapted from *Chemical Engineering*, p. 132, September 1980.

10.12. Maria is a process engineer for Stardust Chemical Corporation, and she has signed a secrecy agreement with the firm that prohibits her from divulging information that the company considers proprietary. Stardust has developed an adaptation of a standard piece of equipment that is highly efficient for cooling viscous plastics slurry. (Stardust decides not to patent the idea but to keep it a trade secret.) Eventually, Maria leaves Stardust and goes to work for a candy-processing company that is not in any way in competition with her former employer. She soon realizes that a modification similar to Stardust's trade secret could be applied to a different machine used for cooling fudge, and at once she has the change made.

A. Has Maria acted unethically (because she divulged the proprietary modification that Stardust developed)?

B. Has Maria acted ethically (because she has used only the idea behind the modification, not the specific change developed by Stardust)?

C. Would Maria have acted unethically if the machine used to cool fudge and the one used by Stardust were identical?

Use one or more of the following heuristics—the four virtues, four-component model, evaluation checklist, ethics checklist, and/or the five P's—to help you answer these questions.

From *Chemical Engineering*, p. 132, September 1980.

10.13. You put together a buoyancy exhibit for a "science day" event. Your exhibit consists of filling up an aquarium with water and then testing which soda pop cans float. The idea is that the diet soda pop cans will float because they don't contain sugar, whereas the non-diet soda pop cans will sink. In preparing your experiment the night before, you discover that one of your diet soda pop cans floats but the other slowly sinks to the bottom. You know that if you "spike" the water with salt, thereby increasing the water's density, all of the diet soda pop cans will float.

What should you do?

A. Spike the water and tell no one?

B. Spike the water and tell only those parents who appear to be looking at your exhibit as a future science fair experiment?

C. Spike the water, but put up a sign next to the exhibit warning people about what you have done?

D. Not spike the water?

E. Do something else? (If so, what?)

Use one or more of the following heuristics—the four virtues, four-component model, evaluation checklist, ethics checklist, and/or the five P's—to help you analyze each of these options.

Problem contributed by Dr. Susan Montgomery, University of Michigan, 1992.

10.14. You attend your friend's piano recital, knowing that your friend is seriously considering a career as a pianist. Her performance is atrocious. She approaches you after the performance and asks, "Well, what did you think?"

What should you do?

A. Tell her to keep her day job?

B. Tell her it sounded terrific?

C. Try to get out of it by telling her a professional couldn't have done a better job?

D. Tell her it was pretty good, with plans to approach her later to discourage her from making a career of it?

E. Do something else? (If so, what?)

Use the four virtues, four-component model, evaluation checklist, ethics checklist, and five P's to help you analyze each of these options.

Problem contributed by Dr. Susan Montgomery, University of Michigan, 1992.

10.15. Your company sends you to a foreign country to take part in the bidding for a large construction project. Once you get there, your associates in the country tell you the bribing rates for the officials in charge of taking the bids.

What should you do?

A. Refuse to bribe the officials, knowing you will certainly lose the project?

B. Bribe the officials at double the ongoing rate, to ensure that you get the project?

C. Go along with the bribing scheme, knowing that it is the only way your project will be considered?

D. Do something else? (If so, what?)

Use one or more of the following heuristics—the four virtues, four-component model, evaluation checklist, ethics checklist, and five P's—to help you analyze each of these options.

Problem contributed by Dr. Susan Montgomery, University of Michigan, 1992.

10.16. Your boss takes you out on the town dining the first weekend of your summer job, not knowing that you are still under the legal drinking age. In the restaurant, everyone at your table orders a drink.

What should you do?

A. Order a beer, wanting to fit in, and hope the server doesn't check your ID?

B. Order a beer, wanting to fit in, and be ready with your false ID in case the server asks for one?

C. Do something else? (If so, what?)

Use one or more of the following heuristics—the four virtues, four-component model, evaluation checklist, ethics checklist, and five P's—to help you analyze each of these options.

Problem contributed by Dr. Susan Montgomery, University of Michigan, 1992.

10.17. Buffalo Nation is a small Native American tribe whose reservation is located in southwestern Colorado. It has three tribal communities on its reservation. Buffalo Nation has debated the feasibility of producing natural gas from the Buffalo Nation Gas Reservoir for more than five years. The tribal council has been split on the issue, however. Younger tribal elders view it as the best way to revive the tribe, as the venture will create significant annual income that will improve the schools, sanitation systems, and roads that have decayed over time, and provide much-needed jobs and income. Traditional tribal elders are against any type of development because the reservoir is adjacent to and under the sacred Buffalo Lake. They view development as invasive to their traditional way of life and as a threat to the tribe's cohesiveness. In the past two years, the tribal council has added several elders who are progressive, including one who has a degree in chemical engineering, and replacing those elders who have become ill or had passed away. (More information—including site proposals and evaluation criteria—can be found on the Web site in the Chapter 10 Professional Reference Shelf.)

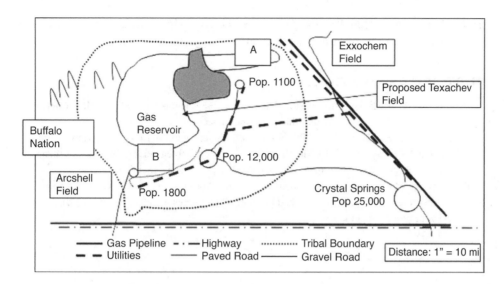

A committee has been formed to reach an intermediate decision about the natural gas proposals. The committee members include the following persons:

– A TexaChev lawyer who has a B.S. in chemical engineering and wants site B to be developed
– The Exxochem field manager, who has a B.S. in chemical engineering and wants site A to be developed
– A tribal lawyer who is a well-respected traditional member from the northern town, is considered to be in line to become the next tribal council leader, and is reluctant to establish a pattern of change for his tribe

– A tribal elder who is a moderate from the central town with education (optional)
– A tribal elder who is a younger member from the southern town and has a B.S. in chemical engineering (optional)

Assume one of the roles of the committee members, and discuss the following issues: evaluation criteria, impact of developing utilities and roads, economic impact, impact of developing a new gas pipeline, environmental impact, cultural impact, impact of potential growth of site, educational impact, and other concerns. *You must discuss each issue from the point of view of your assumed character.* For instance, the industry representatives will favor one of the sites because of the advantages to one particular company. Determine whether a decision can be made to bring the decision before the full Buffalo Nation tribal council.

Material provided by University of Colorado faculty and staff Beverly Louie (Women in Engineering Program/Chemical & Biological Engineering Department) and David Aragon (Multicultural Engineering Program), and Thomas Abeyta of Oberlin College, who developed these materials for use in student leadership development activities.

10.18. Add to the analysis of (a) the "Holiday Gift" case study and (b) the "Rigging the Bidding" case study by applying Seebauer and Barry's four-component model and four virtues.

10.19. Write a paragraph describing the difficulties you might encounter in applying the ethics principles discussed in this chapter on the job in industry.

FURTHER READING

Dresner, M. R. "Risk Assessment: How Do We (Irrational Humans) Really Do It?" *Chemtech*, 21, p. 340, 1991. Provides an introduction to the subject.

Matley, Jay, and Richard Greene. "Ethics of Health, Safety, and Environment: 'What's Right?'" *Chemical Engineering*, p. 40, March 1987, and p. 119, September 28, 1987. More scenarios with a health/safety/environmental flavor to further challenge your ethical judgment.

Rosenzweig, Mark, and Charles Butcher. "Should You Use That Knowledge?" *Chemical Engineering Progress*, April 1992 and October 1992. Interesting ethical scenarios that further test your "ethical judgment."

11 TROUBLESHOOTING*

Previous chapters in this book discussed each of the five building blocks of our problem-solving heuristic. In this chapter we expand on the problem definition techniques introduced in Chapters 3, 5, and 8 to present troubleshooting, another approach commonly used in industry for finding *the root cause* of a problem. Because troubleshooting is frequently applied to technical situations, some of the examples offered here will have a more technical flavor than some of our earlier examples.

   Troubleshooting is used to find out why something is not working as it should be working. It could be that something new, recently constructed, or implemented will not work or it could be that an existing plan or process was working well but then suddenly changed so that it no longer works up to specifications. Troubleshooting can be applied to problems you face in everyday life where you seek to find the real reason why something/someone is not performing as expected. In each of these situations, we need to carry out a methodical analysis to find and correct the problem.

   Troubleshooting is an interactive process in which we form a number of hypotheses about where the fault lies and then carry out procedures to test those hypotheses. We can then systematically eliminate or confirm our hypotheses as we obtain the results of the test procedures we proposed. While troubleshooting is far from an exact science, the guidelines and heuristics we have studied so far in this text (e.g., critical thinking, the Duncker diagram, and K.T. problem analysis)[1] will prove extremely useful in the process.

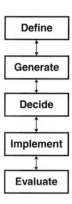

SOME GENERAL GUIDELINES

Successful troubleshooting starts with a solid understanding of the information available, the underlying principles, the process, and the specific unit in question.[1] It also requires paying attention to detail, developing good listening skills, viewing the problem firsthand, and understanding the symptoms of the problem. These and other troubleshooting guidelines are summarized by Laird et al.[2] and shown next.

Troubleshooting Guidelines[2]

1. Gather information.

2. Apply solid fundamental principles.

3. Separate observations from hypotheses or conjectures.

Continues

---

* This chapter's use of troubleshooting algorithm and guidlines is copyright Kepner–Tregoe, Inc., 1994. Reprinted with permission.

4. Independently verify data using field measurements and observations, when possible.

5. Make rigorous comparisons of the data obtained under normal operation with the data obtained under faulty operating conditions.

6. Spend time in the unit making direct observations—even if you are not sure what to expect.

7. Consider the entire system related to the problem.

8. Practice good listening skills.

9. Do not reject serendipitous results.

10. Brainstorm all the things that could explain the fault.

11. Use K.T. analysis (either a problem analysis or potential problem analysis in modified form) and other troubleshooting strategies to deduce what happened during the faulty run. Present the analysis in the form of a table or chart.

12. Do not fall in love with a hypothesis—seek to reject as well as to accept.

13. Suggest a new troubleshooting scenario and again go through the above list.

The first step, gathering relevant information, is a key component in any troubleshooting process. Learn how to ask the critical questions. See if you have the necessary information to make a "ballpark" calculation using the Fermi analysis discussed in Chapter 2. Most engineers at some time or another will be faced with troubleshooting a problem in the plant or manufacturing operation in which they work. You can gather information using the techniques in Chapter 4 such as walking around the plant and talking to George and the operators. To compare data from the malfunctioning unit with data from the normal operation, use the K.T. approach:

- What is? What is not? What is the distinction?
- Where is? Where is not? What is the distinction?

The Kepner–Tregoe (K.T.) approach and the troubleshooting guidelines from Don Woods's book *Successful Troubleshooting for Process Engineers*[3] should be used whenever possible. An excellent set of "rules of thumb" for troubleshooting can be found in this book, along with guidelines for data gathering and developing critical thinking and interpersonal skills. The critical thinking and troubleshooting focus may then be applied to laboratory experiments and field and plant equipment, as demonstrated by the following troubleshooting worksheet developed by Woods.[4] We will also show how the Troubleshooting Worksheet can be used to solve non-technical problems.

Troubleshooting Worksheet[4]

What is the problem? _____

_____

What are the symptoms? _____

    1. _____

    2. _____

    3. _____

Who are the people you will talk to, and why do you want to talk to them? ___

_____

_____

Which data are to be double-checked for accuracy? _____

_____

Fundamentals

What are the guiding principles and equations? _____

_____

List at least five working hypotheses as to the problem:

    1. _____

    2. _____

    3. _____

    4. _____

    5. _____

Monitoring

If I make this measurement or take this action, what will it tell me?

Measurement/Action_____ Reason/Possible Cause _____

Measurement/Action_____ Reason/Possible Cause _____

Measurement/Action_____ Reason/Possible Cause _____

Measurement/Action_____ Reason/Possible Cause _____

Does it Fit the Observation?

Cause of the Problem	Result of the Cause	Does It Fit the Observation?	Steps Needed to Check Cause	Feasibility
1. _____	1. _____	1. _____	1. _____	1. _____
2. _____	2. _____	2. _____	2. _____	2. _____
3. _____	3. _____	3. _____	3. _____	3. _____

Kepner–Tregoe Problem Analysis

What *is*?	What *is not*?	What is the distinction?
1._____	1._____	1._____
2._____	2._____	2._____

Where *is it*?	Where *is it not*?	What is the distinction?
1._____	1._____	1._____
2._____	2._____	2._____

When did it occur (*is*)?	When was everything okay (*is not*)?	What is the distinction?
1._____	1._____	1._____
2._____	2._____	2._____

What *is* the extent?	What *is not* the extent?	What is the distinction?
1._____	1._____	1._____
2._____	2._____	2._____

Which hypotheses are consistent with all symptoms?

_____,_____,_____,_____

Troubleshooters should always keep four or five working hypotheses in mind as they seek to determine what could be causing the fault as they complete this worksheet. Woods stresses the importance of brainstorming as a technique to generate a number of potential explanations. Three of these techniques for brainstorming (e.g., SCAMPER, de Bono's lateral thinking) are discussed in Chapter 7.

Another table that is useful for screening potential causes of a problem is the "Does it fit the observation?" table, which was also developed by Woods:

Cause of the Problem	Result of the Cause	Does It Fit the Observation and/or Measurement?	Steps Needed to Check Cause	Feasibility

This table can help you organize your thoughts by eliminating hypotheses that are not the true cause of the problem.

We will now apply the procedure outlined on Woods's troubleshooting worksheet to three examples: one nontechnical; one requiring very little, if any, of a technical background; and one requiring some understanding of the principles commonly discussed in engineering courses.

Troubleshooting: The Morning Newspaper (Based on a True Story)

You and your wife have just finished unpacking the last boxes from your move last week with your family into your new apartment two weeks earlier. The move has been more exhausting and intense than you planned because you had to rush to make use of the opportunity to move in 10 days earlier than the scheduled move of October 1 because of the early departure of the previous tenants.

During this unpacking, one thing you always could count on was your relaxing mornings with the newspaper and a cup of coffee at the kitchen table. The newspaper is usually delivered by 6 A.M., which is the time your wife leaves for work. A couple of days after you moved in you went downstairs a few minutes before 6 A.M. just in time to see the paperboy drop off the newspaper on the front steps of your two-story apartment building. For the next several days the newspaper arrived at 6 A.M., giving you plenty of time to read it before you needed to leave for work at 7 A.M. However, the past few days it has not been there when you go down to get it at 6 A.M. or even at 7 A.M. when you leave for work. In fact it was not there even when you got home from work. There had been no delivery issues during the first 10 days of living at the new apartment complex, and you recall that you have paid for another six months of subscription. You conclude something must be wrong with the delivery system, and you decide to apply the troubleshooting process to find out what could be behind the missing morning newspaper.

What is the problem?

You stopped getting the morning newspaper two weeks after moving to your new apartment.

What are the symptoms?

There is no newspaper in the morning at 6 a.m. in the front lobby of your building. It was there for the first 10 days.

Who are the people you will talk to, and why do you want to talk to them?

1. Check with your wife to see if she had noticed anything unusual or taken the newspaper to work the few days she had to leave before you got up.
2. Talk to the neighbors to determine if you are the only one experiencing this problem or if others are as well.
3. Contact the paper delivery boy to see if he has been sick.
4. Contact the newspaper to notify them of the problem and ask them if they have stopped delivering your newspaper in the morning for some reason.

Which are the data to be double-checked for accuracy?

Make sure that the online payment for the subscription of the newspaper has been received and that the problem is not caused by an unpaid subscription. Check that your delivery address is the correct one or if an error was made when you went online to change to the new address.

Fundamentals

What are the guiding principles and equations?

When you sign up for a newspaper you fill in the forms online with all your information: name, phone number, address, payment plan, and so on. Each day the delivery boy will then deliver the newspaper by 6 a.m. to your given address in the main hall downstairs in the lobby of your apartment complex.

List at least five working hypotheses for the problem

1. Somebody is stealing or is taking the morning newspaper by mistake.
2. Your online address change never went through or was entered incorrectly.
3. The newspaper is having a problem with its delivery system so it has not been able to send out any newspapers to its customers this week.
4. Your online payment never went through.
5. Your wife is taking the newspaper to work to read and forgot to tell you about it.

Monitoring	
If I make this measurement or take this action, what will it tell me?	
Measurement: Check your online payments in your bank account.	**Reason:** To verify that you are still signed up for a subscription.
Measurement: Check to see that the new subscription address is on file.	**Reason:** To see the changes have been made from the previous address.
Action: Talk to your wife.	**Reason:** To ask if she picked it before you and placed the morning paper inside.
Action: Talk to the neighbors.	**Reason:** To see if you were the only one experiencing this problem.
Action: Call the newspaper delivery boy.	**Reason:** To learn if he has been ill and to inquire why the paper suddenly stopped.
Action: Call the newspaper circulation department.	**Reason:** To inform them about the problem.
Action: Wake up really early and wait until the newspaper arrives.	**Reason:** To catch the paper burglar in act or to see the paper thrown in the bushes.

We now continue our troubleshooting analysis by filling out a K.T. problem analysis table.

K.T. Problem Analysis of the Missing Morning Paper

	Is	Is Not	Distinction	Possible Cause
What (*Identify*)	The morning newspaper all of a sudden stopped arriving.	Receiving the morning newspaper.	The morning newspaper is missing.	The newspaper has stopped sending the morning newspaper to your address; someone is stealing the newspaper or taking it by mistake.
Where (*Locate*)	At your new apartment	At your old apartment	You have a new address.	Did not change the addresses correctly.

Continues

K.T. Problem Analysis of the Missing Morning Paper *(Continued)*

	Is	Is Not	Distinction	Possible Cause
When *(Timing)*	Lately, this last week.	Previous 10 days	You and your family moved.	The address change has not occurred and you have been receiving the former owners' morning newspaper.
Extent *(Magnitude)*	Only your apartment	The neighbors' apartments	There is something wrong with either your payment or address change.	Someone in your apartment complex decided to try your newspaper and took your paper.

Does It Fit the Observations?

Cause	Result	Does It Fit the Observation or Measurement?	Steps Needed to Check Cause	Feasibility
The newspaper has stopped sending the morning paper to you because you stopped paying for it.	Not receiving the morning paper.	Yes. The newspaper might give a few days' grace period then stop the delivery.	Verify payment.	Easy. Log on to your bank account and your newspaper's Web page to see if your subscription is paid.
Did not change the address correctly.	The newspaper is shipped to your previous address.	Yes/No. Yes, the paper would be delivered to another apartment and you would not be receiving it. No, it does not explain you receiving it the first 10 days.	Check the delivery address information on your newspaper's Web page to see if it matches the address at which you are currently living.	Easy. Log on to your account on the newspaper's Web site.

Cause	Result	Does It Fit the Observation or Measurement?	Steps Needed to Check Cause	Feasibility
You moved in early and have been receiving the former owner's morning newspaper during this 10-day period.	You are not aware that it is not your newspaper you have been reading for the past week.	Yes. The previous tenant's payment to the end of the month has now run out and your payment or address change has not been received and therefore you did not receive your paper.	Contact and talk to the former owner.	Easy or difficult depending on forwarding information of previous tenant?
Someone stole or took your morning newspaper by mistake, including family members.	Not receiving the morning paper.	Yes, someone could be stealing your paper.	Wake up before 6 a.m. and meet the delivery boy to ensure that no one else is able to take your newspaper before you.	Easy

Results of the Steps Checked to See If the Hypothesis Fit the Observation

On checking with the bank and with the paper it was found that the bill was paid, refuting hypothesis number 4. However, when checking with the office that delivers the papers to investigate hypothesis number 3, it turned out that when you and your family moved, you went online and entered your new address on your account on the newspaper's Web site but forgot to hit "SUBMIT" at the very end to change the delivery address to the new one. During the first 10 days in the new apartment, which were the last 10 days of the month, you were receiving the newspaper from the last person staying in your apartment that had paid for delivery through the end of the month and not updated his delivery address. When the month ended and he finally changed his information it appeared as if your morning newspaper was not being delivered but in fact you had been getting his newspaper the first 10 days and yours was still going to your old address. You finalize the address change on the newspaper's Web site and begin to receive your newspaper.

In this troubleshooting example, we used the troubleshooting procedure to solve a generic problem. The remaining examples in this chapter apply the troubleshooting process to problems of an engineering nature.

TECHNICAL TROUBLESHOOTING EXERCISES

Troubleshooting: Gold Production

In the mining industry, gold and other metals are extracted from rocks by adding chemicals in a process called leaching, which is represented in the following figure. In the extraction stage (1) a cyanide solution is mixed with rocks, allowing the cyanide solution to bind to the gold ions in the rocks, thereby extracting the gold from the rest of the rock. The gold is then precipitated from solution and the mixture is fed into the separation stage (2), where the gold particles are removed from the wastewater solution containing the cyanide and leftover rock. The wastewater stream exiting Stage 2, and fed to the storage tank (3), contains large amounts of cyanide and other environmentally hazardous metals that must be treated before being discharged into the environment. The wastewater stream from the storage tank (Stage 3) is passed through several cleaning mixers in order to achieve sufficient water purity (Stage 4). In the cleaning stage, large plastic bags of chemicals are added manually to the wastewater. The hazardous compounds in the wastewater react with the added chemicals and form sludge, in a process called flocculation. The sludge sinks and is removed from the bottom of the cleaning tank, decreasing the toxicity of the system, and clean water flows out of the top of the tank.

FrameAngel/Shutterstock

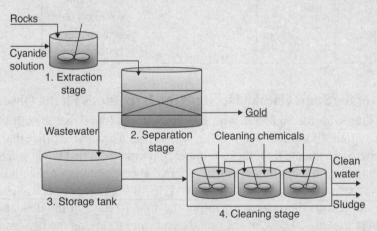

Schematic of Gold Digger's gold production facility (developed with Nikola Popara).

Gold Digger is one of the world's leading gold-producing companies. Its facility was designed and operated according to the previously described process represented by the figure.

Last week, Gold Digger received notification from the Environmental Protection Agency (EPA) that the discharged water from the plant had unacceptable and dangerously high levels of cyanide and metals, leaving Gold Digger susceptible to heavy penalties and perhaps even a lawsuit. The equipment technician is new and inexperienced and has not been able to determine the cause of the problem. He states that he has been following standard operation procedure by throwing the plastic bags of flocculating chemicals into the mixers every day. Gold Digger's management team now turns to you to investigate why the discharged water stream after the cleaning stage is no longer clean enough and to determine why this decrease in water quality occurred.

Troubleshooting Worksheet

What is the problem?

The EPA says that the discharged water from the plant has unacceptably high levels of cyanide and metals and wants the problem solved immediately.

What are the symptoms?

Measurements confirm cyanide and metals are present in the water stream leaving the plant.

Who are the people you will talk to, and why do you want to talk to them?

- The process engineers who designed the process can provide information about how the process works and how the different stages connect with one another.
- The equipment technicians who monitor the equipment and are responsible for the different treatment stages can provide information about any changes that have occurred with the process.
- The laboratory technicians who conduct all of the analysis and keep track of all the data of the process (e.g., concentrations, temperatures, etc.) can provide information to determine if there are any irregularities.

Which data are to be double-checked for accuracy?

- The water stream after the cleaning stage, Stage 4, needs to be rechecked to confirm that the exit water stream is indeed contaminated and does not meet the EPA levels.
- The instruments that measure cyanide and metals concentrations need to be recalibrated to make sure the concentrations are indeed above EPA limits.
- The concentrations of cyanide and other hazardous metals must be measured to determine the severity of the situation.
- The concentration of the different chemicals added to the solution in the cleaning stages needs to be measured.
- The inside of the storage tank should be checked to see if any irregularities have occurred.

Continues

Fundamentals

What are the guiding principles and equations?

Cyanide solution is being used to extract gold from rocks. The wastewater from the extraction is separated from the gold and then the wastewater must be processed. The cleaning is achieved by adding a chemical that will flocculate the contaminants, leaving a sludge and clean water. For the process to work effectively, the chemical flocculant must be completely dissolved in the wastewater.

List of working hypotheses for the problem:

1. There is leakage in the cleaning stage, leading to sludge escaping with the water stream.

2. The amount of chemicals added to the cleaning stage was insufficient to flocculate the particles and form a sludge, causing different metals and small amounts of cyanide to remain in the system.

3. The instrument that measures cyanide and metals concentrations is not properly calibrated and not yielding accurate results.

4. The process technician that operates the cleaning stage could have made an error in adding the flocculating agent and other chemicals, and they did not get dissolved adequately in the wastewater.

5. The mixing in the cleaning stage is too fast, breaking up the flocculated aggregates and not allowing the contaminants to properly settle to the bottom and be separated.

Monitoring	
If I make this measurement or take this action, what will it tell me?	
Measurement: Metal concentration in the solution in each stage and tank.	**Reason:** Determine if any stages or tanks have high metal concentration. A high metal concentration indicates that the water is contaminated.
Measurement: Concentration of added flocculating chemicals in the cleaning stage.	**Reason:** An insufficient amount of chemicals will not form the sludge to remove all of the contaminants.
Action: Analyze a sample of the contaminated water that is entering and leaving the system.	**Reason:** To identify precisely what is contaminating the water.
Action: Recalibrate the instrument that made the measurements of the contaminated water.	**Reason:** An uncalibrated instrument would yield the wrong results.

Action: Check the mixing speed of the cleaning stage.	Reason: Excessive mixing can make it hard for the particles to sink to the bottom, leading to insufficient cleaning.

K.T. Problem Analysis of the Dirty Water Stream

	Is	Is Not	Distinction	Possible Cause
What	Particles (cyanide and metals) present in the discharged water stream	Uncontaminated discharge stream	Particles are present in the discharged stream that should not be there	Not proper operating conditions (temperature, pH, mixing speed, or concentration of the added chemical flocculant).
Where	The stream after the cleaning stage	All other streams in the facility	The liquid stream is the only flow out of the cleaning stage. All other waste is removed as solid.	The cleaning stage equipment has not been operating correctly or is leaking.
When	Recently	A few months ago	A new equipment technician was hired.	The new equipment technician was not very familiar with the procedure.
Extent	Only in the cleaning stage	Any other stages	Only the cleaning stage removes the particulates.	Ineffective mixing of flocculant or the flocculating chemicals are not added correctly.

Continues

	Unacceptably high concentrations of particles in the water stream	Low concentration or no particles in the outlet stream	Particle size. Larger aggregates of flocculated particles will sink faster.	Inefficient flocculation and removal.

Does It Fit the Observations?

Cause	Result	Does It Fit the Observation or Measurement?	Steps Needed to Check Cause	Feasibility
Too high mixing speed in the cleaning stage.	Creates turbulence in the tank, which makes it harder for the formed particles to remain flocculated and sink to the bottom and be separated.	Yes, without good flocculation the particles would not be removed in the cleaning stage.	Open the tank and observe the mixing.	Easy—Vary the mixing speed and measure the water quality again.
Insufficient amount of flocculating chemicals added to the cleaning stage.	Poor sludge formation and flocculation of particles, which leads to poor cleaning.	Yes, the lack of cleaning chemicals would allow the water to pass through the cleaning with more particles than planned.	Take a sample of the water from the bottom of the cleaning stage.	Easy—Add a new round of chemicals and observe.
The instrument that measures the water quality is not operating correctly.	Yields incorrect values of the water quality.	Yes, the incorrect readings of clean water would make the EPA think the plant is producing water that does not meet specification for discharge when indeed it does.	Recalibrate the instrument and run a trial.	Easy—Follow a standard procedure.

Cause	Result	Does It Fit the Observation or Measurement?	Steps Needed to Check Cause	Feasibility
Leakage in the cleaning stage	A leak could cause precipitated particles to escape with the water stream.	Yes, if the precipitated particles are removed with the clean water stream the water will still be considered contaminated.	Visual observation.	Easy

Results of the Steps Needed to Check the Different Causes

The equipment technician measured the mixing speed and it was the same as before the decrease in water quality occurred, which removed hypothesis 5 from contention. Before the EPA notification, laboratory results showed extremely low values of the chemicals in the liquid sample withdrawn from the tank. Laboratory technicians reported after recalibrating the instrument that the results were the same as before, therefore refuting hypothesis 3. A thorough visual inspection of the equipment was conducted and no traces of any leakage could be found, falsifying hypothesis 1.

However, when the investigators added a new round of chemicals, the problem disappeared. It was found that the new technician did not open the plastic bags with the cleaning chemicals before adding them in to the cleaning stage; he simply threw the bags over the side of the mixer and into the solution in the cleaning tanks. As a result the chemicals never entered the solution, confirming hypotheses 2 and 4: that a human error occurred and an insufficient amount of chemicals were added.

This example also points out the need to make sure the operators have an understanding of the process and are properly trained. The technician did not know it was the contents inside the bag that had to be mixed into the solution, and he just followed his instructions to add the bags to the tank without question. Because the plastic bags were sealed and did not dissolve in water, no chemicals were released and the sludge was never separated. By opening the bags and pouring the chemicals into the tank the problem was easily solved and saved Gold Digger a significant amount of money in not having to pay any fines.

In the next example, we will recall some of the principles of heat transfer and apply them to an actual case history from Marlin and Woods. This example is included for those students who are either enrolled in engineering or are thinking of enrolling in engineering.

Troubleshooting: The Boiler Feedwater Heater

Waste flash steam from the ethyl acetate plant is saturated at slightly above atmospheric pressure. It is sent to the shell and tube heat exchanger to preheat the boiler feedwater to 70°C for the nearby boiler house. The boiler feedwater heater is shown in the figure. Condensate is withdrawn through a thermodynamic steam trap at the bottom of the shell. The water flows once through ¾-inch nominal tubes. There are 1000 tubes. "When the system was put into operation three hours ago, everything worked fine," says the supervisor. "But now the exit boiler feedwater is 42°C instead of the design value. What do we do? This problem is costing us extra fuel to vaporize the water at the boiler." Fix the problem.

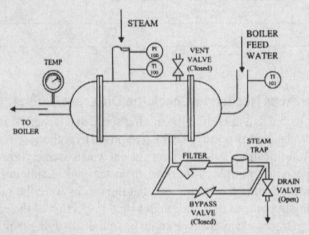

Source: Example of In-Class Exercise from Marlin/Woods Case History. Case 28 from D. R. Woods, *Successful Troubleshooting for Process Engineers*, Wiley-VCH Verlag, Darmstadt, Germany, 2006.

Troubleshooting Worksheet

What is the problem?

Heat exchanger malfunction.

What are the symptoms?

The water temperature at the exit of the heat exchanger is 42°C instead of 70°C, resulting in boiler feedwater not being heated to the appropriate temperature.

Who are the people you will talk to, and why do you want to talk to them?

The plant engineer will know how the equipment functions. He says this is the first time workers have used waste steam from the ethyl acetate plant.

Which data are to be double-checked for accuracy?

The temperature of water from the heat exchanger has been wrong in the past. It was double-checked and found to be 42°C.

Fundamentals

What are the guiding principles and equations?

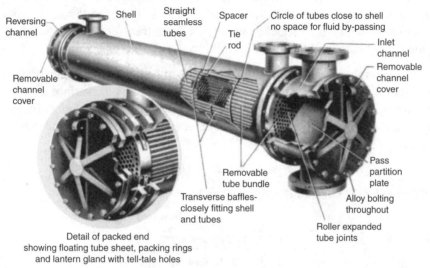

*Two-Pass Shell-and-Tube Heat Exchanger Showing
Construction Details*
Source: Courtesy of Ross Heat Exchanger Division of American Standard. From
Max S. Peters, Elementary Chemical Engineering, 2nd ed., McGraw-Hill,
New York, 1984.[4]

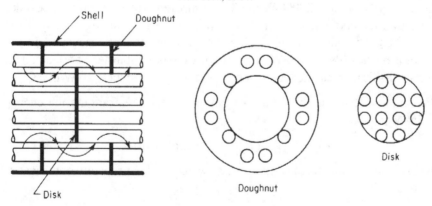

Disk-and-Doughnut Baffles
Source: From Max S. Peters, Elementary Chemical Engineering, 2nd ed.,
McGraw-Hill, New York, 1984.[5]

The water being heated flows on the inside of the tubes, and the waste steam condenses on the outside of the tubes. The rate of heat transfer, Q, from the

Continues

condensing steam at temperature T_{steam} to the water at temperature T_{water} at any point in the exchanger is given by the equation

$$Q = UA\,(T_{steam} - T_{water})$$

The heat flux at any point is $Q = UA(T_{steam} - T_{water})$, where T_{water} is the local temperature. The energy balance is $Q = UA\Box T_{lm}$. This equation for Q using the log mean temperature driving force may be rearranged to the equation for $T_{out,}$ where A is the heat exchange area and U is the overall heat transfer coefficient, which is a measure of the rate of heat transfer between the shell-side fluid (steam) and the tube-side fluid (water).

The smaller the value of U, the lower the rate of heat transfer. Neglecting the resistance of the pipe wall and assuming that the inside pipe surface area and the outside pipe surface area are the same, the overall heat transfer coefficient U is related to the individual heat transfer coefficients inside (h_i) and outside (h_o) the tubes by the equation

$$1/U = 1/h_i + 1/h_o$$

The smaller the value of the h's, the smaller the value of U.

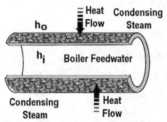

For condensing steam, the outside (i.e., shell-side) heat transfer coefficient is large—on the order of 20000 W/m²/°C—compared to the inside (i.e., tube-side) heat transfer coefficient for flowing water (which is approximately 1500 W/m²/°C). The heat transfer coefficient for stagnant water surrounding the pipes is on the order of 100 W/m²/°C. If air were present instead of steam, then the shell-side heat transfer coefficient would be approximately 10 W/m²/°C.

We know that we can find the exit temperature by rearranging the energy balance to obtain

$$T_{out} = T_{steam} - (T_{steam} - T_{in})\exp(-UA/mC_p)$$

where m and C_p are the mass flow rate of water and heat capacity of the water, respectively.

Consequently, we see that the outlet temperature from the exchanger to the boiler, T_{out}, would be lowered if the inlet and steam temperatures were lower than expected; the mass flow rate, m, were higher than expected; the overall heat transfer coefficient, U, were lower than expected; and the effective heat exchange area, A, were smaller than expected. The overall heat transfer coefficient will be lower if the individual transfer coefficients inside (h_i) and outside (h_o) were smaller than expected.

List at least five working hypotheses for the problem:

1. The steam trap has become blocked, causing liquid condensate to back up in the heat exchanger and preventing the steam from contacting the pipes in the exchanger.

2. The entering water is subcooled.

3. The steam pressure and the temperature have dropped.

4. The heat exchanger has become fouled.

5. The waste steam from the ethyl acetate plant contains noncondensable gases.

Monitoring	
If I make this measurement or take this action, what will it tell me?	
Measurement: Inlet temperature.	**Reason:** Subcooled inlet.
Measurement: Water flow rate.	**Reason:** A higher than normal flow rate could cause the fluid not to reach 70°C.
Action: Check whether the steam trap is closed and not functioning properly. If it is functioning correctly, it should open and close periodically as condensate is formed in the shell.	**Reason:** Water may be filling up the shell side of the exchanger, reducing the condensing steam heat transfer area, resulting in a lower overall heat transfer coefficient that is much smaller than that of condensing steam.
Action: Check whether the filter is plugged.	**Reason:** This problem would produce the same symptoms as a closed steam trap.
Action: Carefully open the vent.	**Reason:** If noncondensable gases have accumulated in the shell, the steam-side heat transfer coefficient, h_o, would be significantly decreased, reducing U.
Action: Make sure the drain valve is open.	**Reason:** If someone has closed the drain valve, water may be filling up the shell side of the exchanger, reducing the condensing steam heat transfer area and overall coefficient.
Action: Check the inlet steam temperature and pressure.	**Reason:** If either of these measurements has decreased, the enthalpy of the entering steam will be less than expected, reducing the outlet water temperature.

We now continue troubleshooting analysis by filling out a K.T. problem analysis table.

K.T. Problem Analysis of the Boiler Feed Heater Problem

	Is	Is Not	Distinction	Possible Cause
What	Low exit temperature	Normal or too-high exit temperature	Insufficient heat supply to raise the temperature to 70°C	Flow rate is too high and inlet water temperature is too low, leading to insufficient heating rate of the fluid.
	Correct water and steam temperature measurements	Wrong temperature measurements	Current temperature driving force should be sufficient to heat the water to 70°C because condensing steam has a high heat transfer coefficient	Something other than incorrect measurements
	Normal water feed rate	High water feed rate	Current temperature driving force should be sufficient to heat the normal water flow to 70°F	Something other than a high flow rate
	Filter and steam trap open	Blocked filter or steam trap	Tubes not surrounded by liquid condensate	Something other than liquid is increasing the resistance
When	Three hours after start-up	Immediately after start-up	Decrease in heat transfer	Buildup of noncondensable gas from waste steam or shell filled with water

	Is	Is Not	Distinction	Possible Cause
Where	Inside heat exchanger	Outside heat exchanger	Entering temperatures are normal	Inefficient heat transfer between shell and tubes
	Entering steam and water temperature normal	Abnormal entering water or steam temperatures	Temperature driving force is not affected	Heat transfer resistance is increased
Extent	Only part of the equipment is affected; some tubes are not affected	All of the equipment is affected; some tubes are not affected	Heat exchange takes place between the shell and tubes	Inefficient heat transfer between the shell and tubes

Does It Fit the Observation?

Cause	Result	Does It Fit the Observation or Measurement?	Steps Needed to Check Cause	Feasibility
Fouling/scale on water side or on steam side	Decrease in heat transfer coefficient	Does not account for a temperature drop over a *short period*	Instrumentation and measurements to calculate heat transfer coefficient; inspection of tubes	Inspection of the tubes is time-consuming and costly if instruments are not available
Malfunctioning steam trap; clogged condensate valve	Rise in water level and consequent loss of condensing steam heat transfer area	Water buildup could account for the temperature drop	Observation of water level in condenser shell	Easy—shut down condenser and remove drain
High presence of inert elements in steam; clogged bleed valve	Decrease in heat transfer area and shell-side coefficient	Inert element buildup may account for temperature drop	Shut down, vent inert elements, and restart the system; perform a bleed gas analysis	Availability of skilled technician and equipment?

Continues

Does It Fit the Observation? (*Continued*)

Inaccurate temperature reading	No actual malfunction in the boiler	Does not account for a temperature drop over a short period	—	—
Steam super-heated too high; water flow too high	No malfunction	Does not account for a temperature drop over a short period or a large temperature drop	—	—
Drop in steam pressure owing to a steam-side leak	Change of condensation temperature	No visible/audible signs of a steam leak	—	—

Source: Example of In-Class Exercise from Marlin/Woods Case History, Case 28 from D. R. Woods, *Successful Troubleshooting for Process Engineers*, Wiley-VCH Verlag, Darmstadt, Germany, 2006.

Results of the Steps Needed to Check the Different Causes

The condenser was shut down and both the stream and condensate valve were found to be fine, refuting hypothesis number 1. Then the whole system was shut down, inert elements vented, and a bleed gas analysis was conducted. It turned out that "dirty steam" containing noncondensable gases had blanketed the heat exchanger tubes in the shell of the heat exchanger. Steams sometimes can contain other gases (most frequently air) that will not condense at the temperatures encountered in the condenser. Since the air does not condense, it tends to build up on the steam side of the exchanger, and it negatively affects the performance of the condenser. The air effectively blankets some of the surface area of the tubes. Consequently, the condensing steam could not effectively reach the tubes carrying the cooling water. With the noncondensable gases gone, the system functioned properly so further tests into other hypotheses were not necessary.

INTERACTIVE COMPUTING MODULE ON TROUBLESHOOTING

You can also hone your troubleshooting skills by using the interactive computer module (ICM) on troubleshooting included on the Web site (www.umich.edu/~scps) that accompanies this book. A number of ICMs have been developed for chemical reaction engineering (CRE)[6] and for problem solving, and they are available from the CACHE Corporation. These modules have been enthusiastically received by many students who have used them in the past.[7,8]

In the ICM on troubleshooting, you are asked to troubleshoot a micro plant that manufactures styrene from ethyl benzene and is not operating properly.

As shown in the diagram, the plant consists of two pre-heaters to the reactor, a reactor, a condenser, a liquid–gas separator, a liquid–liquid separator, an adsorption system, and a distillation column.

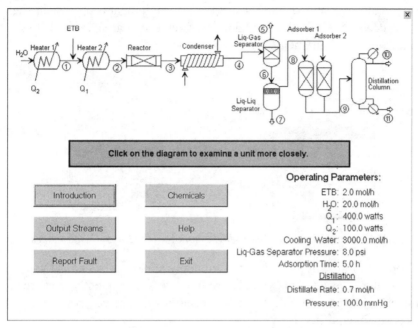

Screen Shot of the Micro Plant

There are a number of potential faults in each unit, any one of which could be causing a problem. Two faults are chosen randomly each time you log on to the ICM. After you log on to the ICM, you can access two sets of simulated data. One set gives instrument measurements such as temperatures, pressures, and flow rates for a number of the streams for normal operation. The other set gives the same readings for faulty operation.

					Component Mole Fractions					
Values for Stream 5										
	Flowrate (gmol/hr)	Temp (deg C)	H2O	ETB	STYR	BENZ	TOLN	METH	ETHL	H2
Expected	0.506	63.016	0.000	0.006	0.945	0.023	0.001	0.013	0.000	0.011
Actual	0.164	63.016	0.000	0.005	0.967	0.015	0.000	0.004	0.000	0.008

Done

Comparison of Actual Measurements and Expected Measurements for Stream 5

You must find the two faults by interacting with the computer to take measurements and obtain the operating conditions of each piece of equipment. Each time you request information from the computer, you are charged a specified amount of money, depending on the complexity of the request, as you would be in real-life situations. You can choose one or more pieces of equipment on which to make measurements and obtain the results. A list of instrument readings and measurements typically available on the chosen piece of equipment is provided to help you make this decision. The following figure uses the reactor as an example to show the type and cost of measurements that can be carried out.

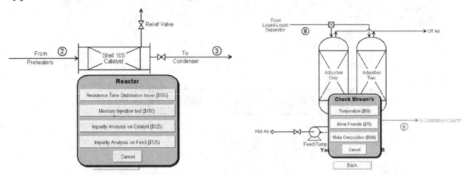

Measurements Available to Make on the Reactor and Adsorbers

You are limited in the amount of money you have to spend, so you must be prudent in the measurements you choose to make. Consequently, you are encouraged to use the same troubleshooting procedures discussed previously to target your measurements effectively.

TROUBLESHOOTING LABORATORY EQUIPMENT

When one of the authors of this book (Scott Fogler) teaches the senior unit operations laboratory at the University of Michigan's College of Engineering, the last three weeks of laboratory sessions are devoted to having students apply their newly learned troubleshooting skills. In these sessions, students work in groups of three on one of four different pieces of equipment: a packed bed distillation column, a double-effect evaporator, an Advanced Reactor Safety Screening Tool (ARSST), and a Continuous Stirred Tank Reactor/Plug Flow Reactor (CSTR/PFR) apparatus. Before students try to carry out an experiment on these pieces of equipment, the graduate student instructors (GSIs) generate a specific fault in each one of these pieces of equipment and collect the data. The students are then given two sets of data. One set is taken when the equipment was operated under normal conditions. The second set of data includes the same measurements for the same equipment, but this time they were obtained under the faulty operation planned by the GSIs.

After the students familiarize themselves with their particular piece of assigned equipment they are asked to perform the troubleshooting algorithm and worksheet shown on page 273. During their troubleshooting exercises, they are allowed to submit three questions in writing to the professor/GSI regarding the data supplied to them. These troubleshooting exercises on real equipment provide students with some very valuable hands-on experience that serves them very well when they become practicing engineers in industry.

SUMMARY

Troubleshooting is used to determine why something new, just constructed, or recently implemented will not work. It is also applicable to an existing plan or process that was working well but then suddenly changed so that it no longer works up to specifications. In each of these situations, we need to carry out a methodical analysis to find and correct the problem. Troubleshooting is an interactive process—as demonstrated by Woods's troubleshooting worksheet—in which we form a number of hypotheses about where the fault lies and then carry out procedures to test our hypotheses. We can then systematically eliminate or confirm our hypotheses as we obtain the results of the test procedures we proposed, thereby identifying the root cause of the problem.

WEB-SITE MATERIAL (WWW.UMICH.EDU/~SCPS)

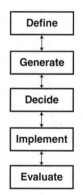

- **Learning Resources**

- **Professional Reference Shelf**
  Interactive Computer Modules—Troubleshooting Exercise

REFERENCES

1. Kepner, C. H., and B. B. Tregoe, *The New Rational Manager*, Princeton Research Press, Princeton, NJ, 1981.
2. Adapted from Laird, D., B. Albert, C. Steiner, and D. Little, "Take a Hands-on Approach to Refining Troubleshooting," *Chemical Engineering Progress*, p. 68, June 2002.
3. Slightly modified from Worksheet 2-1 from D. R. Woods, *Successful Troubleshooting for Process Engineers*, Wiley-VCH Verlag, Darmstadt, Germany, 2006.
4. After Donald Woods, McMaster University, Hamilton, Ontario, Canada.
5. Peters, Max S., *Elementary Chemical Engineering*, 2nd ed., McGraw-Hill, New York, 1984.
6. Fogler, H. S., *The Elements of Chemical Reaction Engineering*, 4th ed., Prentice Hall, Upper Saddle River, NJ, 2006.
7. Fogler, H. S., S. E. LeBlanc, and S. M. Montgomery, "Interactive Creative Problem Solving," *Computer Applications in Engineering Education*, 4, 1, pp. 35–39, 1996.
8. Fogler, H. S., and N. Varde, "Asynchronous Learning of Chemical Reaction Engineering," *Chemical Engineering Education*, 35, p. 290, 2001.

EXERCISES

11.1. Load the micro plant simulation from the Web site (www.umich.edu/~scps). Carry out an analysis to learn the two faults in the plant. (*Note:* The faults are randomly chosen so that each time you sign on there will be two different faults.) Turn in your analysis and your performance number to your instructor.

Performance number:_____

11.2. "Early Morning Shivers": This exercise is an introduction to troubleshooting on your own. In the following example, many parts of the troubleshooting process have already been completed. Fill in the missing information as noted by "_____" to complete the example problem.

Your spouse shakes you awake at 5 a.m. on a late January morning. The house is cold and it is snowing outside. You usually turn the thermostat back a bit at night, but the temperature in the house is 52°F (you determine this fact by looking at your desk thermometer). Something is obviously wrong. Your spouse says the house is too cold for your 18-month-old daughter and asks you to do something to get the heat back on. As a temporary measure, your spouse brings your daughter into your bed to keep her warm (temporarily solving one "real" problem). You start to solve the problem by mentally filling out the troubleshooting worksheet.

Troubleshooting Worksheet

What is the problem? _____

What are the symptoms? _____

Who are the people you will talk to, and why do you want to talk to them? _____

Which data are to be double-checked for accuracy? _____

Fundamentals

What are the guiding principles and equations?

The thermostat (located in the living room) controls the furnace, which in turn heats the house through the combustion of natural gas in the furnace. The flame heats a heat exchanger through which the air passes before it is circulated throughout the house. When the temperature in the house falls below the desired value (the set point), the thermostat calls on the furnace to start. When the house warms up to the desired temperature, the thermostat turns the furnace off. The furnace incorporates several safety devices.

- A high-temperature sensor shuts off the natural gas flow if the temperature inside the furnace gets too high.

- A flame sensor detects whether the pilot light is lit. This very small flame burns continuously and ignites the main gas flow. If the pilot is off, it would be unsafe to open the gas valve, because unburned natural gas would then flow into the furnace and the house, creating an explosion hazard. (Some newer furnaces have a glow plug that heats up to ignite the gas, but the principle is the same: If the glow plug fails to heat up, the safety system prevents the main gas valve from opening.)

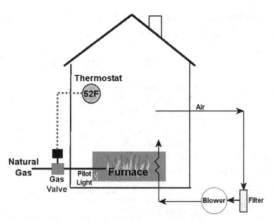

List at least five working hypotheses for the problem:

1. The thermostat is malfunctioning and the furnace does not "think" the house is cold.

2. The electricity in the house is off, or the furnace circuit breaker has tripped.

3. The natural gas supply to the house has been interrupted.

4. The furnace air filter is plugged.

5. The pilot light is out.

6. The sensor for the pilot light is dirty and doesn't sense that the pilot light is on.

Monitoring	
If I make this measurement or take this action, what will it tell me?	
Measurement: _____ _____	**Reason:** A dead battery can cause a malfunction.
Measurement: _____ _____	**Reason:** If the furnace water is hot, the natural gas supply to the house is okay, because the water heater also operates on natural gas.
Action: _____ _____	**Reason:** This action confirms that there is not an electrical power outage to the house, which would affect the furnace.
Action: _____ _____	**Reason:** If the furnace breaker is tripped, the furnace will not operate.
Action: _____ _____	**Reason:** A plugged or dirty filter can cause the furnace to not operate properly.

Continues

Continue your troubleshooting analysis by filling out a K.T. problem analysis table and a "Does It Fit the Observation?" table.

	Is	Is Not	Distinction	Possible Cause
What	House is cold	Normal or too high temperature	_____ _____ _____	Furnace is off
When	Sometime since bedtime last night	Prior to bedtime	_____ _____ _____	Low outside temperature
Where	Entire house is cold	Localized to one room	_____ _____ _____	Thermostat problems; natural gas outage; plugged furnace filter
Extent	Only furnace	Water heater or other utilities	_____ _____	Thermostat problems; plugged furnace filter; other, more complicated mechanical/ electrical problems

Does It Fit the Observation?

Cause	Result	Does It Fit the Observation or Measurement?	Steps Needed to Check Cause	Feasibility
Thermostat battery is dead	_____ _____ _____	Yes	Remove and replace battery	Easy, if a spare battery is handy
Furnace filter is blocked or partially blocked	_____ _____ _____	Yes	Remove filter and temporarily run without one	Easy

Cause	Result	Does It Fit the Observation or Measurement?	Steps Needed to Check Cause	Feasibility
Electricity to furnace is out	_____ _____ _____	Yes	Reset circuit breaker	Easy
Natural gas supply to heat the house is shut off	_____ _____ _____	Yes	Check hot water heater	Easy to check—call the gas company
Circuits or connections in furnace are corroded or malfunctioning	_____ _____ _____	Yes	Varies—may need to open furnace up	Need a trained technician

The problem in this example turned out to be a weak thermostat battery. Replacement of the battery corrected the problem.

11.3. Lake Nyos in western Cameroon, adjacent to Nigeria, in the elbow region of West Africa, is a water-filled throat of an old volcano. The lake, which is deep and funnel sloped, lies within the Oku Volcanic Field, at the northern boundary of the Cameroon Volcanic Line. The Cameroon Volcanic Line is a zone of crustal weakness and volcanism that extends to the southwest through the Mount Cameroon stratovolcano. The Oku Volcanic Field contains numerous basaltic scoria cones and maars. Lake Nyos itself occupies a maar crater that formed from a hydrovolcanic eruption 400 years ago. Although the volcano is no longer erupting, gas continues to be released very slowly, directly into the deepest water of the lake, by the old plumbing system. This gas is virtually all carbon dioxide, with no traces of deadly hydrogen cyanide or hydrogen sulfide. There are about 30 similar lakes in the region.

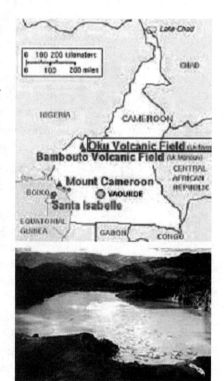

Lake Nyos covers an area of about 1.5 square kilometers and is more than 200 meters deep. The region of western Cameroon where it is located averages about 2.5 meters of rain each year. In the rainy season, the excess lake water escapes over a low spillway cut into the northern rim of the maar crater and races down a valley toward Nyos village.

On the morning of August 13, 1986, approximately 1700 people in the villages of Nyos, Kam Cha, and Subum, and virtually all the cattle, birds, and other animals living along the shore of Lake Nyos and in the valley were found dead. The hill people who discovered the disaster found some people still in their beds, some dead on the floor in their houses, and some dead along the water's edge.

The people who lived in the hills above the lake—all of whom survived—recalled hearing a bang around 9:30 P.M. the night before. It was neither a volcanic eruption nor an earthquake, and there were no visible signs of any destruction to the houses or to the trees and other foliage. However, for several days before August 12, there had been a number of unusually cold, rainy days. As a consequence, the cold water ran down the sides of the volcano into the lake, which was at a higher temperature. The water in Lake Nyos is normally a beautiful, deep-blue color. However, the lake now appears to be composed of murky, reddish brown water that apparently formed by the oxidation of iron-rich bottom waters that were carried up to shallower lake levels during the August 1986 event.

There is a rumor that a similar tragedy befell the inhabitants around Lake Nyos more than 100 or 200 years ago.

Source: www.geology.sdsu.edu/how_volcanoes_work/Nyos.html.

Use one or more troubleshooting techniques to find the cause of this disaster. After finding the cause, carry out a potential problem analysis to suggest ways to prevent such a disaster from occurring again.

One way to use this troubleshooting exercise in an interactive manner is to write down a list of questions to give to your instructor to answer. Your instructor, who has the solution, will charge a specified number of points for each question you ask; the point value (from 5 to 25 points) charged will depend on the quality of the question and the effort needed to find the answer. You have 100 points to spend, so you will not want to ask all of your questions at once but rather wait until you receive an answer before posing the next question. You want to spend as few points as possible. Each of the following sites may help you ask a pertinent question:

> www-personal.umich.edu/~gwk
> www-personal.umich.edu/~gwk/research/nyos.html

11.4. Photolithography is a process used to make the computer chips that are found in many electronic devices, including cell phones, MP3 players, digital cameras, and laptop computers. Photolithography uses light to transfer patterns from a "master" (i.e., a mask) to a surface (e.g., a silicon wafer). These patterns ultimately become the electrical components (transistors, resistors, wires, and so forth) of the chip. The pattern of light is generated by shining light through a mask, which selectively blocks the light in a manner similar to an overhead transparency. The light pattern is focused onto a substrate coated with a thin polymeric film called a photoresist. Acid is generated in the regions of the film that are exposed to light. The acid catalyzes a chemical reaction, which in turn allows the exposed regions of the protective film to become soluble. The soluble regions can then be washed away using a "developer" fluid, leaving the unexposed regions intact on the substrate. The exposed areas are further processed by depositing metal, etching, doping, and other means.

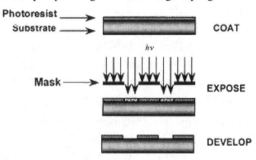

The photoresist typically contains a polymer and a photo-acid generator (PAG). An example of a PAG is shown below. The presence of light causes the molecule to break apart and form an acid (H+).

Ideally, the photogenerated acid reacts with a component of the polymeric resist that changes the solubility of the resist. This change in solubility is the critical step, allowing exposed regions of the resist to be dissolved away when placed in a developer. A "deprotection" reaction changes a portion of the polymer coating, allowing it to be dissolved away so that the area is no longer "protected." An example of a deprotection reaction is shown here.

Protective Coating   *Deprotected Coating (Can Be Dissolved)*

The acid acts as a catalyst—it is regenerated during the reaction. A single acid can catalyze many deprotection reactions. This approach to patterning is called "chemically amplified resists," because a single photon is capable of ultimately deprotecting many sites on the polymer by generating a single acid molecule. The ultimate effect is to make the resist (i.e., the protective coating) more sensitive to light, which improves throughput.

Company X has been successfully using the photolithography process described previously for more than a year. Unfortunately, at the end of February the plant's East Wing began producing faulty chips. The faulty production of chips continued through the first few weeks of March. The source of the trouble appears to be that the exposed regions of the chips produced in the East Wing are not all dissolving as expected, causing the resulting chips to malfunction. Operations in the West Wing are unaffected. Although this problem has been occurring only on Friday afternoons for the past few weeks, it is critical that it be solved.

When you gather additional information about the situation, you make the following discoveries. Only one worker handles the chips in each wing of the plant. In the East Wing, the chips are handled by Mary, who has been employed since the beginning of the year. John, the West Wing operator, has been doing this job for several years now. You decide to observe the workers on Friday. The morning seems to go normally, and there are no problems with the process. You decide to have lunch with John and Mary. The company cafeteria changed caterers at the beginning of the year and the food is quite good, with a broad menu capable of satisfying the diverse work force at the plant: dishes ranging from tuna fish casserole (beneficial for reducing stomach acidity) to Asian cashew chicken (beneficial for increasing protein intake) to beef and potatoes (beneficial for people with traditional diets and tastes). John remarks how he really enjoys the new steak sandwiches and suggests that you try one. You do and are impressed with the quality. During lunch, conversation focuses on a wide variety of topics, from company politics, to gorging on too many paczkis on Fat Tuesday last week, to sports (March Madness is in full swing and Mary is a big basketball fan), to Mary's upcoming trip to celebrate Easter with her family in New York. Lunch passes uneventfully, and the afternoon shift begins—and so do the problems with the chips. What do you conclude?

One way to use this troubleshooting exercise in an interactive manner is to write down a list of questions to give to your instructor to answer. Your instructor, who has the solution, will charge a specified number of points for each question you ask, and the point value (from 5 to 25 points) charged will depend on the quality of the question and the effort needed to find the answer. You have 100 points to spend, so you will not want to ask all of your questions at once but rather wait until you receive an answer before posing the next question. You want to spend as few points as possible.

Problem courtesy of Professor Michael Dickey, Department of Chemical Engineering, North Carolina State University.

11.5. "Dow Corning's Product XYZ": Dow Corning Corporation is one of the global leaders in silicones and silicon-based production technology. All products are produced under the same production procedures and in the same process equipment. Dow Corning Corporation uses only two raw materials, A and B, for all their products, varying the feed ratios to create different products. In between each batch the equipment is cleaned with a solvent to make sure the next batch has the correct ratio of A and B.

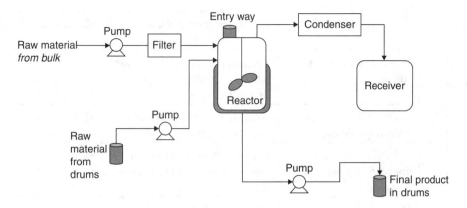

Raw material A arrives in bulk containers and is filtered while being pumped into the process in order to remove impurities. Raw material B arrives prefiltered in drums from another production site and is pumped directly into the process. The process consists of a reactor, a condenser, and receivers. The receivers collect the liquid that condensed in the condenser in order to safely dispose them as waste. Gaseous ammonia is used to pressurize the reactor, and it is supplied from a bulk storage tank. The final product is then pumped to drums for use in other processes.

Dow Corning's product XYZ is a clear, highly viscous material that is manufactured in this process. The internal customer (i.e., another process that uses a product from one process as a raw material in another process) reported that it had found black specks in the bottom of a drum of product XYZ that had been manufactured three months prior. Another batch of XYZ was produced with new drums and examined to see if it was just the previous batch or if it was the drums that were filthy and caused the contamination. But this result yielded the same levels of black specks, confirming that product XYZ, not the drums, was contaminated with black specks. Because product XYZ is highly viscous, the problem could not be easily solved by filtration.

Dow Corning has had some complaints in the past regarding impurities with its output streams. One case was when the condenser's cooling system was not operating as designed, which led to an insufficient cooling and poor cleaning. Not long after that incident, the reactor caused serious problems, with leaking valves and malfunctioning pumps. Dow solved this problem by replacing the old-fashioned valves and pumps in the process with new high-performance ones. A year ago, Dow Corning experienced another problem with the reactor, but this time it was the hatch at the top of the reactor that caused problems.

The hatch's gasket was made out of Teflon and had cracked, because it was opening and closing too many times, and started to leak out. The company fixed this minor problem by simply replacing the Teflon gasket with a stronger graphite gasket.

Your job is to find out what is causing the contamination problem and fix it. Your project budget is $10,000 for investigating the operation of the plant and maintenance and to perform tests and analysis. The remaining money, once the case is solved, will be your profit.

	Questions	Measurements and Actions	Cost to Perform
1.	What are these black specks?	Perform analysis to determine what contaminants are present in the drums.	$1500
2.	Do the raw materials contain particles?	Analyze both raw material B and raw material A.	$2500
3.	Is the filtration unit operating properly as designed?	Check the filtration unit for raw material A by performing tests.	$1000
4.	Do the pumps operate as designed?	Check the pump units.	$1000
5.	Do the condenser and receivers operate as designed?	Check the condenser and receiver units.	$3000
6.	Is the reactor operating as designed?	Examine the maintenance of the reactor.	$6000
7.	Is there something leaking into the reactor?	Open the hatch and visually observe if anything had been dropped in the reactor.	$2000
8.	Is enough gaseous ammonia supplied to the reactor in order to create optimum conditions?	Check the flow rate.	$1000
9.	Are the drums filthy or could the area around the drums have caused the contamination?	Examine the areas around where the material is put into drums.	$500

Problem and photo courtesy of Maria Quigley, Dow Corning.

FURTHER READING

Woods, Donald. *Rules of Thumb for Engineering Practice*. Wiley-VCH, Weinheim, Germany, 2007.

12   PUTTING IT ALL TOGETHER

In the previous chapters of this book, we presented the building blocks of creative problem solving individually so that we could focus more easily on each block. Of course, when we are faced with a problem in the real world, we bring many of these principles together to bear on the problem at the same time. We have shown the problem-solving heuristic in the form of a job card here. Job cards are usually displayed at the workplace or carried around by individuals to help them recall previously learned concepts that can be applied to their jobs.

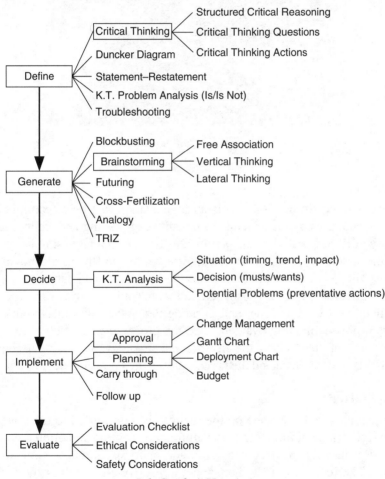

Job Card: A Heuristic

We will now show how the principles we have been studying can be applied to two case studies.

CASE STUDY 1: GENERATING SUSTAINABLE POWER FOR AN AGRICULTURAL PROCESSING PLANT IN INDIA

This case study was contributed by Marina Miletic, Ph.D., while in the Department of Chemical and Biomolecular Engineering, University of Illinois, Urbana–Champaign, Illinois.

The EWB Team in India

Background

Engineers Without Borders (EWB) is an organization that helps communities in developing countries by introducing sustainable engineering and technology. The University of Illinois's student chapter of EWB received a request to help create a sustainable source of power for an off-the-grid village in Bhadakamandara, India. One of the village's sources of income was the production and harvesting of spices and other agricultural goods. These materials were sold to a buyer who further processed them by drying, milling, and grinding the spices and other products. The members of this community wanted to do the processing themselves to generate more income, but did not have the electricity, technology, or training needed to do so. The EWB students accepted the task of starting this power generation project.

Getting Started

This project required everyone on the team to be committed to the goal and to be in the right frame of mind for both solving the problem and implementing the solution. Because this project was voluntary, many participants came and went over time. The few who made the project a priority did much of the planning, went to India, worked until the process was implemented, and then worked to further improve the process even after the project was complete. All students worked equitably toward the goal, and everyone had well-defined roles and objectives throughout the entire project. These students were so focused on the end goal that few personal conflicts arose. The interactions of the group exemplified the team skills discussed in Chapter 2, as evidenced by everyone listening to others' ideas and respecting others' efforts.

There were numerous opportunities for solving many highly important problems, such as the need to purify the unclean water in the area and the need to deliver power to everyone's home in the village. Because of time and resource constraints, however, the problem was limited to providing power for agricultural processing.

Problem Definition

Gathering Information

Data Collection. To further define this problem, data needed to be collected. In particular, the following questions needed to be answered:

1. What kind of power generator will we use?
2. What kind of fuel will it use? (What kind of fuel is naturally occurring and/or available in that area?)
3. Where will the power generator and other equipment be housed?
4. How will we construct the building that will contain the generator and the other processing equipment?
5. How will we obtain the necessary supplies in India?
6. How will we communicate with the local residents? Where do we get an interpreter?

Talk with People Who Are Familiar with the Problem. EWB has undertaken many projects in India and around the world. The University of Illinois chapter of EWB submitted its proposal to work on an international project to EWB-USA headquarters (located in Colorado). The national headquarters office partnered the university team with the Association for Indians' Development (AID-Orissa), with Orissa University of Agriculture and Technology (OUAT), and later with Jagannath Institute of Technology and Management (JITM). All of these organizations had unique information to contribute to help identify the problem.

View the Problem Firsthand. A group of students performed a site assessment to ask local residents about available resources. The team found a vendor for generators in India and identified areas where they could buy related supplies. The team (after talking with villagers) decided on the site for the building where the generator and other equipment would be stored. This information was integrated into the next step of the process.

Problem Definition Techniques

The following critical thinking actions, which were discussed in Chapter 3, helped the students define the problem.

Predicting. In predicting, we envision a plan and its consequences. The team had anticipated some of the major challenges involved in this project, both technical and social. This anticipation helped the group with planning a timeline and assigning tasks so that problems were resolved according to schedule.

Analyzing. In analyzing, we separate or break the whole into parts to discover their nature, function, and relationships. By first narrowing the problem and then breaking the problem down into parts, the EWB team was able to identify the main obstacles for each problem. This narrowing helped identify the resources that were needed (people, time, and money) to complete the project.

Information Seeking. In information seeking, we search for evidence, facts, or knowledge by identifying relevant sources and gathering objective, subjective, historical, and current data from those sources. By speaking with EWB headquarters, AID-Orissa, OUAT, JITM, and local residents, the team was better able to determine which kind of generator and agricultural equipment would be needed and where it would be assembled. They also learned about the potential health and cultural problems they might encounter.

Applying Standards. In applying standards, we make judgments based on established personal, professional, or social rules or criteria. The team agreed that they would go ahead with the project as long as what they did benefited the villagers. Safety was a top priority. These standards were maintained throughout the process and were incorporated in all decisions.

Problem Statement
Ultimately, the students defined their problem as follows: "Generate sustainable power for running agricultural processing equipment."

Generating Solutions
To generate a solution to this problem and answer the questions posed earlier, the team did a lot of brainstorming and research. Many technical blocks had to be removed before implementation of a solution to achieve the goal of powering the process.

Several possibilities for types of generators and fuels were identified. The team quickly identified biofuel- or vegetable oil–run generators as some of the best options given the size and scale of the spice-processing equipment. Ideas regarding the construction materials of the building were generated. Because the building would be constructed on a rice paddy, the students had to think about the materials for the foundation as well as for the building itself. A number of types of building materials were considered (including brick and concrete) based on availability.

Deciding the Course of Action
Students first distinguished which problems needed to be solved immediately and which problems could wait until later. Furthermore, some aspects of the problem had a greater priority than others. The timing of deciding the type of generator and

developing a design for it had the highest impact and the highest timing priority. Other aspects of the problem, such as setting up grinding and milling machines, could be dealt with later. The following K.T. situation appraisal (KTSA) chart identifies the students' project priorities.

Situation Appraisal*

Project	Timing	Trend	Impact
1. Decide on type of generator and fuel	H	H	H
2. Design generator and fuel system for the specific process	H	H	H
3. Set up grinding and milling machines	L	M	H
4. Purchase auxiliary supplies	L	M	M

Because of the remote location, the students decided to use a proven technology to drive the power generator—specifically, an internal combustion engine. Two viable choices were then identified for internal combustion engine power generators: one that used biodiesel as the fuel and one that used vegetable oil as the fuel. There were several critical criteria that the system had to fulfill, and the solution was identified based on these requirements. Although the formal problem analysis (KTPA), decision analysis (KTDA), and potential problem analysis (KTPPA) processes were not used explicitly, some systematic approaches were used to converge on the best solution. The following KTDA is a representation of the process of converging on the chosen solution.

Decision Analysis

	Running on Vegetable Oil	Running on Biodiesel
Musts		
Generator must be available for purchase in India	Go	Go
Generator must use renewable fuel as power source	Go	Go
Generator must use fuel that is available locally	Go	No Go
Generator must use fuel that is safe (no toxic chemicals)	Go	No Go

---

* Appraisal, decision analysis, potential problem analysis, and problem analysis charts in this chapter are copyright Kepner–Tregoe, Inc., 1994. Reprinted with permission.

Wants	Weight	Rating	Score	Rating	Score
System should be easy to maintain	9	6	54	X	X
System should be low in cost	7	5	35	X	X
System should last a long time	6	5	30	X	X
System should not involve too many steps	5	6	30	X	X
Total			**149**		

The biodiesel option was eliminated because it required the use of methanol and sodium hydroxide for fuel production. These components were not readily available in the area and were also considered potentially dangerous. Vegetable oil could be used if an oil press were available to press seeds that grew naturally in the area. A disadvantage of using vegetable oil was that it meant modifying the engine to add a fuel preheater and a separate tank.

The Modified Generator for Use with Vegetable Oil

One major concern for the team was the possibility of getting sick during their one- to two-month stay in India. Analyzing potential problems such as snake bites and other illnesses helped the team determine that they needed to have a driver available at all times to take them to the nearest hospital. The following KTPPA outlines preventive actions and contingency actions for this situation.

Potential Problem Analysis

Potential Problem	Causes	Preventive Action	Contingency Action	Triggers
Becoming sick in India.	Drinking the unclean water.	Bring lots of water filters.	Determine the problem's severity and seek medical attention. Use the driver as needed.	Symptoms such as diarrhea, vomiting, and stomach pain.
	Getting bitten by a snake.	Research what kinds of snakes are in the area and what to do in case of a bite.	Seek medical attention. Take antivenom. Use the driver as needed.	Contact with snake, penetration of skin.

Implementing the Solution

The first step to implementing the solution was the approval process. Approval was necessary from EWB-USA as well as from the various nongovernmental organizations with which the team was working. Most importantly, the villagers had to agree to the solution proposed and the construction process, which would require time and space for its completion. This approval process required substantial communication between all groups. It was only after this approval was granted that implementation could be started.

After identifying the solution and receiving approval, the team set its priorities, identified numerous tasks, and assigned roles to each individual on the team. Responsibilities such as grant writing, communications with local residents, engine design, and process equipment design were broken up between numerous teammates, and deadlines were set. A timeline (Gantt chart) was developed and included in some of the grant proposals the team wrote. Following is an example of a deployment chart, which features the team's responsibilities.

Deployment Chart

Task	Sean Poust	Ben Barnes	Stephanie Bogle	Patrick Walsh
1. Engine design				X
2. Grant writing/ travel	X			
3. Project leader			X	
4. Process assembly	Everyone			
4. Solar dryer design		X		

The team faced numerous obstacles while trying to implement the solution. The site that they thought would hold the building with the generator was different than originally planned. The team members had to build the structure on rice paddy land. They purchased bricks for building the structure that would house the generator and other equipment, only to find when these bricks arrived that many were misshapen or broken, though they were told that the bricks would still work in the structure. After trying to assemble the building, the team quickly realized that the structure was unstable. They sought the advice of an expert and insisted that a civil engineer in charge of one of the nongovernmental organizations make a six-hour drive to the site to inspect the structure. The team ended up wrapping the entire building in wire mesh and concrete to ensure its stability. The team recognized when it needed more help and sought it out.

Inside the Building

The Completed Building

After the generator was set up and mounted to the floor of the building, the team encountered another problem: The generator vibrated excessively. This problem required troubleshooting because the level of vibration was unacceptable and could lead to machine failure. The following KTPA illustrates the what, where, when, and extent of this problem.

Problem Troubleshooting (KTPA): Generator Vibrating Excessively

		Is	Is Not	Distinction	Possible Cause
What	Identify	The generator is vibrating excessively	Any other equipment in the building vibrating	Something specific to the generator	Uncentered motor within generator
Where	Locate	The generator and the pad on which the generator is mounted	Anything else in the building	Only the generator itself or the pad	Poor shock absorption of pad
When	Timing	The problem occurs only when the generator is running and was initially observed when the generator was first turned on	The problem does not occur when the generator is off	Motor motion causes vibration	Mounting pad is not appropriate for motor/ generator
Extent	Magnitude	Large unacceptable vibration	Small or no vibration	The generator and the pad are not compatible	Possibly need a different shock- absorbing pad

The team tried mounting the generator on a different pad, one that featured more shock absorption. This change did not seem to help significantly. The generator was mounted very tightly to the pad, and eventually the team heard a loud cracking noise while the generator was running. A fastener bolt that held the generator to the pad had sheared. This breakage caused the generator to be more free-floating and hence reduced the vibrations significantly. Thus the problem was not with the generator but rather with the way it was mounted to the pad. Once the generator was mounted more loosely on the pad, the vibrations subsided considerably. Consequently, *the problem defined as "how to generate sustainable power" was solved.*

One cultural problem that the team encountered proved difficult to solve, especially because of the language barrier between the students and the local residents. Women were not allowed to be part of the process training, even though it was their idea to start the project in the first place. It was only appropriate for the female members of the EWB team to interact with the village women, so they continued to work on this problem throughout the process. Its solution is still ongoing.

Evaluation

One of the most important aspects of any project is evaluation. The team evaluated its decisions and made corrections at every point along the way. At the end of the project, a final evaluation was completed in which the students determined whether more work needed to be done and which tasks were the highest priorities.

The evaluation of the implementation process can be performed by answering questions from the evaluation checklist.

Does the solution solve the real problem?

The solution does solve the problem originally defined—a lack of sustainable power for running agricultural processing equipment. Although other problems remain, such as the need for better lighting and clean water, this solution does provide the village with a generator that can increase the standard of living for its citizens.

Is the problem permanently solved, or is this a patchwork solution?

Like most solutions, this one will require maintenance and care taking. The most difficult part was setting up the equipment and training local residents to use it. Ensuring that the process will continue to work will require long-term dedication to equipment maintenance, replacement, and operator training.

Have all the consequences of the solution (both negative and positive) been examined?

Despite the benefits, some negative aspects of this solution include finding dedicated people to run the process and sell the product continuously.

The village must maintain a team of people who take responsibility for running the process, making decisions regarding its use, and managing the money generated.

Has the solution accomplished all it could?

The solution might possibly have included providing electricity for the entire village, though this was beyond the scope of the project's timeline and initial resource allotment. The team has been exploring means of providing electricity for these homes as a follow-up project.

Is the solution ethical?

The team considered whether or not the team should have taken a more proactive approach to involve the women in the training. The team decided not to address this issue because it could not be resolved in the short time they were in the village and that as such they might have caused more problems than they solved.

Is the solution economically, environmentally, and politically responsible and safe?

Because the process uses vegetable oil—a renewable resource that is available locally through plants—it is environmentally friendly. The process was designed so that the chemicals used would be biodegradable and not hazardous to human health. Furthermore, the extra income it provides can be used to create a better infrastructure for the villagers so that they can improve their water quality, so long as the money is managed properly.

In the end, the team assembled the building, designed the process, trained local residents to use the equipment, and have continued to monitor the progress of the project. Although communication with this remote village is quite challenging, the team tries to maintain contact with the local people who are now running the process to make sure it is still going and attempts to troubleshoot problems remotely. The team has continued to work on other problems faced by the village. In particular, team members have been pursuing the idea of replacing indoor kerosene lanterns, which often cause asphyxiation and other health problems, with small-scale, solar-powered, battery-operated LED lanterns. This project continues to be an exciting challenge for these engineering students, encouraging them to do well and make the greatest impact in the lives of others.

To learn more about EWB, please visit www.ewb-usa.org.

CASE STUDY 2: PROBLEMS AT THE BAKERY CAFÉ

Project Assignment

This case study is the result of a term project in the senior-level course at the University of Michigan titled "Problem Solving, Troubleshooting, and Making the Transition to the Workplace" (Engineering 405—the course syllabus is included on the Web site (www.umich.edu/~scps)). It was contributed by Carly Ehrenberger, James Herbin, Paul Niezguski, and Sara Schulze, with additional input from Ana Butka, Jessica Moreno, and Jason Rhode-McGauley.

Students were given the following instructions: "Interview employees of Bakery Café on Truman Parkway to determine any problems the employees have. Analyze the problems you uncovered and then generate and evaluate solutions for each problem. Finally, pick the best solution for each problem you identified."

Background

The Bakery Café features fresh food and beverages that are delivered in a comfortable environment for its customers. Customers order at a counter from a diverse menu and wait for a short time while Bakery Café employees prepare their orders. Bakery Café offers its customers amenities including comfortable seating, free wireless internet, and a catering service. The Bakery Café has more than 30 employees who work a variety of shifts and jobs. Typical jobs include working the register, making sandwiches and salads on the "line," washing dishes, assisting in catering, and acting as a trainer or supervisor. The ages of employees range from 15 to 55 years. They each work an average of 30 hours per week and have worked for Bakery Café for an average of 15 months.

Getting Started: Methodology

Using the "five building blocks" problem-solving process, the following steps were taken:

1. Define the problem.
   - Gathered information about the Bakery Café.
   - Developed, distributed, and collected an employee survey after attending a workshop led by a business school professor.
   - Interviewed the manager.
   - Used various problem-solving techniques, such as a Duncker diagram and a Kepner–Tregoe potential problem analysis.
2. Generate solutions.
   - Used the idea generation techniques discussed in Chapter 7 to develop a number of solutions to each problem.

3. Decide on the course of action.
   - Chose the best alternative solution for each problem identified and analyzed.
4. Implement the solution.
   - Recommended future actions for the Ann Arbor store manager to take.
5. Evaluate the solution.
   - Evaluated the feasibility of the solutions recommended.

Problem Definition

Gathering Information

Data Collection

- Researched Bakery Café online to get some background and basic corporate information.

Talk with People Who Are Familiar with the Problem

- *Interviewed manager*: To gain information about the current situation at Bakery Café, we asked critical thinking questions such as "What types of problems do you expect we will find?" to understand the situation at hand and determine the areas of focus.

- *Employee surveys (22 responses)*: To help understand the basic concerns or areas obviously in need of improvement at Bakery Café as well as to uncover underlying or unspoken problems.

View the Problem Firsthand

- Visited Bakery Café for dinner and observed employees and customers.

Confirm All Findings and Continue to Gather Information

- *Searched for biases in survey answers*: Survey results were analyzed in light of each employee's age or position to understand potential biases.

Development of a survey was a key means of gathering the necessary information to define the problem. Brainstorming techniques and Socratic questions were the tools used to help develop the survey. The formulation of the survey was extremely important and required significant thought. Guidelines for survey development were obtained and used from the Web site http://www.statpac.com/online-surveys/index.htm, along with other material given in the references at the end of this chapter.[1–7]

Developing the Survey:
Application of Brainstorming Techniques

Uncovering the Unspoken Problems

We began with the problem statement: "What are possible topics that we should ask employees about that could reveal the unspoken problems with their jobs?"

Application of three techniques—free association, lateral thinking using random stimulation, and analogy and cross-fertilization—helped to generate survey questions. The following questions were developed.

Free Association

Fellow employees:

- "Are your co-workers generally friendly and willing to help you if needed?"
- "Does Bakery Café deal fairly with all employees?"

Managers:

- "Does the manager value the employee's contribution?"

Shifts:

- "Are you satisfied with the number of hours/shifts you work each week?"
- "Is the way the shifts are scheduled convenient?"
- "Is the way the shifts are scheduled flexible?"

Job tasks:

- "Do you know what is expected of you?"

Stress level:

- "Are you under too much stress at work?"
- "Is there a sufficient number of workers present during each shift?"

Pay:

- "Does your pay match your job performance?"

Benefits/rewards:

- "What do you like most about working for Bakery Café?"
- "Are you recognized when you do a good job?"

Lateral Thinking Using Random Stimulation

Random word: **Knife** → cutting bread → task at work

- "What is your favorite task?"
- "What is your least favorite task?"

Random word: **Voice** → freedom of speech

- "Can you voice your opinions without fear of criticism from your manager?"

Analogy and Cross-Fertilization

What could a Bakery Café manager learn from an elementary school teacher?

Elementary school teachers understand that students need time to learn and are patient with them even when they make mistakes.

- "Does your manager use mistakes as positive learning experiences?"

Elementary school teachers give students grades and report cards to show students their progress and to set goals for the future.

- "Do you receive regular feedback that helps improve your performance?"

The Socratic questions discussed in Chapter 3 were also employed to help develop survey questions.

Application of Types of Socratic Questions to Developing the Survey

Clarification: Why do you say that?

- "What is your favorite task to do at work?"
- "Why do you like this task?"
- "What do you like and not like about this task?"
- "What is your least favorite task?"

Probe Reasons and Evidence: What evidence do you have to support your answer?

- "Why do you feel the customers are not being served efficiently?"
- "Why do you feel there might be a problem with shift scheduling?"

Explore Viewpoints and Perspectives: What is a different way to look at it?

- "Which areas need improvement?"
- "What would you suggest doing to improve them?"

After the survey was administered, the responses were examined to determine the problems. A total of 22 surveys from employees were analyzed. The following main concerns were identified:

- Bottlenecks occur because not enough workers are present during each shift.
- Co-workers do not take responsibility for job tasks.
- Employees do not receive enough training for tasks.

Problem Definition Techniques

A Kepner–Tregoe situation appraisal was performed on these perceived problems to determine their priority order and the next steps to be taken in addressing them.

KTSA of Employee Problems at Bakery Café

Problem	Timing	Trend	Impact	Next Process
1. Co-workers do not take responsibility for job tasks.	M	H	H	KTPPA
2. Bottlenecks occur because not enough workers are present during each shift.	M	M	H	KTDA
3. Employees do not receive enough training for tasks.	M	M	H	KTPA

The rankings of each of the three evaluation criteria are based on the benefit to Bakery Café as an organization and to this store location in particular.

Problems 1, 2, and 3 are moderately urgent in the *timing* category. All of them directly impact the customers and ultimately the success of the store.

In the *trend* category, problem 1 was assigned a high degree of concern (H): If co-workers are not taking responsibility for their assigned tasks, then unfinished tasks will accumulate and a hostile environment could be created among employees. Problems 2 and 3 have both been given a ranking of moderate concern (M). Not having enough workers on a shift might continually escalate employee and customer frustration as well as leaving tasks unaccomplished. Similarly, a lack of training might increase employee frustration.

In the *impact* category, problems 1, 2, and 3 were all sources of high degrees of concern because they impact both the employees and the customers. The possible impacts of these problems are as follows:

- Problem 1: Poor cleanliness and general disorganization
- Problem 2: Slow service
- Problem 3: Inconsistent service and/or product

If these problems are solved, the result might potentially create a more harmonious work environment for employees. This could, in turn, create better manager–employee relationships as well as more satisfied customers.

Based on the KTSA, the following priority was given to the problems identified:

- Problem 1: Co-workers do not take responsibility for job tasks.
- Problem 2: Bottlenecks occur because not enough workers are present during each shift.
- Problem 3: Employees do not receive enough training for tasks.

This priority should determine the order in which the problems are addressed.

Generating Solutions and Deciding the Course of Action

The next step in the KTSA was to decide on a course of action. For problem 1, potential problem analysis (KTPPA) was selected because both problems have future implications and their root causes have not been identified.

Problem 1: Co-workers Do Not Take Responsibility for Job Tasks

The KTPPA helped identify possible causes of the problem and a plan to prevent future pitfalls.

KTPPA for Job Task Responsibility

Potential Problem	Possible Causes	Preventive Action	Contingency Plan
Co-workers do not take responsibility for tasks, leaving the work to be performed by workers on later shifts	Not enough time	Create and use stockpiles of prep material	Add another employee
	Lack of awareness that a task needs to be done	Post checklist of duties near the punch-out clock	Manager completes task
	Blatant neglect on the part of some shifts	Reward employees who complete tasks for other shifts	Confront those shift workers who neglect tasks; schedule the most efficient workers for earlier shifts

Problem 2: Bottlenecks Occur Because Not Enough Workers Are Present During Each Shift

For problem 2, decision analysis was selected as the analytical technique because the cause of the problem had already been identified and a course of action needed to be implemented. Problem 2 was particularly evident during morning and late-evening shifts. The survey revealed that some workers had to wait to use the espresso machine while others were moving as fast as they could while taking orders and serving pastries. Customers would complain as the line of customers became longer. The problem statement was formulated as follows: "Choose a way to better serve customers in the mornings and late evenings." Alternative solutions identified by brainstorming included hiring more workers, purchasing additional equipment, transferring workers between shifts, or preparing early for shifts. The *musts* and *wants* for the solution are shown in the following table.

KTDA for Not Enough Workers Present during Each Shift at Bakery Café

		Hire More Workers		Purchase Additional Equipment		Transfer Workers between Shifts		Prepare for Shift Early	
Musts	Decrease customer complaints	Go		Go		Go		Go	
	Increase preparedness between/during shifts	Go		Go		Go		Go	
	Maintain Bakery Café's philosophy of freshness	Go		Go		Go		No Go	
Wants	*Weight*	*Rating*	*Score*	*Rating*	*Score*	*Rating*	*Score*	*Rating*	*Score*
Increase task completion rate	6	5	30	7	42	7	42	X	X
Increase shift satisfaction	9	8	72	5	45	4	36	X	X
Decrease customer wait time	6	5	30	7	42	5	30	X	X
Decrease worker idle time	7	4	28	6	42	6	42	X	X
	Total		**160**		**171**		**150**		

Because the results of the decision analysis were very close, the possible adverse consequences associated with each alternative had to be closely examined.

KTDA: Adverse Consequences of Not Enough Workers per Shift

Alternative	Probability	×	Severity	=	Threat
Hire More Employees					
Increase payroll	10	×	3		30
More difficulty in scheduling shifts	6	×	7		42
Total					**72**
Purchase Additional Equipment					
Maintenance cost	10	×	1		10
Worker training on equipment	7	×	2		14
Total					**24**
Reallocate Worker Distribution					
Reduced shift productivity	7	×	3		21
Scheduling conflicts	6	×	2		12
Reduced employee satisfaction	5	×	6		30
Total					**63**

From the analysis of adverse consequences and results of our KTDA, it appears that Bakery Café should purchase additional equipment to increase the shift productivity, allowing workers to serve customers more quickly during the peak morning and late-evening rushes. Specifically, surveys revealed that a bottleneck was created when many customers ordered espressos. Therefore, we suggested that Bakery Café purchase an additional espresso machine identical to the one it currently operates. This purchase would increase productivity without increasing the amount of training required or person-hours logged. Additionally, reallocation of workers to some of the busier shifts (perhaps with some employees working occasional split shifts) would help to alleviate this problem.

Problem 3: Employees Do Not Receive Enough Training for Tasks

Our surveys revealed that employees felt there was a lack of training among their fellow employees. This notion was not limited to employees—managers expressed this perception as well.

Before analyzing this problem and finding ways to solve it, we wanted to verify that the root problem of employee competencies was the employee-training program. Triggers from the problem statement–restatement technique, which was discussed in Chapter 5, were applied to redefine the problem once more. For this technique, a slightly different original problem statement was used: "Some employees believe other employees do not always perform well on the job."

Trigger 1. With this trigger, an emphasis is put on different words or phrases in the original problem statement. The following sentences emphasized different parts of the problem statement to make us think of a new problem statement that will be more tightly defined than the original.

1. Some **employees** believe other employees do not always perform well on the job.

   In statement 1, the word "employees" was emphasized, which led to the question of whether only the employees noticed the poor performance of the other employees or whether customers noticed the problem as well. This trigger also raised the question of whether the manager noticed any poor performances by some of the employees. If this was the case, did the manager ever make sure that all of the employees had the proper information needed to perform well on the job?

2. Some employees believe **other employees** do not always perform well on the job.

   Statement 2 placed emphasis on "other employees." This statement questioned the relationships between the employees, including whether they were all on amicable terms with one another. Delving deeper into the statement recalled a few of the survey comments that complimented the "less-than-par" employees on their personalities and ability to get along, yet criticized their work performance.

3. Some employees believe other employees do not always **perform well** on the job.

   The third statement placed the emphasis on "perform well," which questioned the standards of performance on the job. How were these standards made, and who made them? These questions led back to the employee-training problem: If all of the employees were trained to a specific standard, then would any of them show a significant lack of performance?

4. Some employees believe other employees do not always perform well **on the job.**

    In statement 4, emphasis was placed on the phrase "on the job." This part of the sentence raised the question of what the employees were like outside of the work environment. If the employees were dependable, hardworking people in other aspects of their lives, then they were probably likely to show the same characteristics while working at Bakery Café. An interview before a new employee was hired could reveal whether the interviewee had the necessary traits for working at Bakery Café.

Placing emphasis on different parts of the sentence gave us some ideas about how to redefine the problem. We continued using the problem statement triggers to see whether the problem could be focused more narrowly.

Trigger 2. Trigger 2 substitutes explicit definitions into the original problem statement. The original problem statement was rewritten as follows:

> "Some people who work at Bakery Café indicated on the survey that other people who work at Bakery Café do not complete all of their assigned job-related tasks while they are working at Bakery Café."

Restating the problem this way led to the question of whether the employees know all of the "assigned job-related tasks"—that is, their work requirements. This led back to the problem in employee training, which, if revised, could make sure all employees know the requirements of working at Bakery Café.

Trigger 3. Trigger 3 was applied to learn whether the training program was the real problem or whether a different problem was at fault. Trigger 3 reworded the original problem statement to say the opposite of what it should. The following opposite statement was found to be quite helpful:

> "What can be done so that all employees feel that all other employees never perform well on the job?"

This opposite statement raised the question of what would cause an employee to always perform poorly. If all employees were trained very poorly, they would probably always perform poorly.

Using these three triggers, the original problem statement was redefined as follows: "Find a way to make sure all employees are well trained and competent."

Duncker Diagram

Once this problem statement was created, we devised the following Duncker diagram:

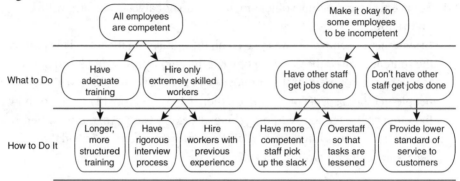

Duncker Diagram for Employee Competence

Generating Solutions by Free Association

We then brainstormed ways to improve employee training by using free association. The following alternatives were identified:

- *Practice runs before opening.* The new hire would arrive before the store opens to practice daily tasks such as working the register and making sandwiches. A manager or qualified trainer would supervise these practice runs, and the new hire could become acquainted with various tasks without the added element of customer service.
- *Employee tests.* The new hire would be required to pass tests on all common tasks before completing his or her training. These tests would be monitored by a manager or qualified trainer to ensure that the new hire is capable in all areas before training completion.
- *Specific number of days training.* All new hires would be required to complete a specified number of days in training. Their shifts would also have to include a certain number of day and night shifts to guarantee that the trainee would be capable in all areas. Based on personal experiences and suggestions from current employees, the belief is that training during three opening shifts and three closing shifts would provide the new hire with the best training experience.
- *Training with a manager.* All training shifts would need to be supervised by a manager. Having a manager present at all times would help with the overall consistency of the training because only one or two people would ever be in charge. The trainee would also be less likely to develop poor habits because the manager is probably the most aware of proper work procedures.
- *Make sure employees are trained on all tasks.* Employees would need to spend time training on all tasks. The person in charge of training could

develop a checklist to guarantee that each task is addressed and that the trainee is competent in each area.

- *Employees specifically assigned as trainers.* Having one or more employees specifically assigned as trainers could improve the consistency of the training. The trainers would need to be qualified as knowing the proper procedures for all tasks. They would also need to know the standard training procedure to be followed in all training sessions. This procedure would not only help to make sure that a competent employee is training the new hire, but would also relieve the training pressure on the manager.
- *Computer simulation with built-in testing.* A computer simulation could be built to represent the daily tasks and challenges that an employee might encounter. Quizzes or other interactive activities could be built into the simulation to confirm that the trainees are retaining the information they learn.
- *Employee worksheets with menu and standard procedures for each shift.* New hires would be required to complete worksheets with questions based on the menu and standard procedures for each shift, including opening and closing.
- *Work instructions present and in plain view at each workstation.* Small cards with proper work instructions would be placed in appropriate areas around the restaurant. These work instructions would need to be clear and concise to ensure that the new hire is easily reminded of the tasks to be performed at each station without the instructions becoming a distraction to employees and customers.

Deciding the Course of Action

Next, these alternatives were examined using a decision analysis to determine the best solution to the training problem. Using a situation appraisal to rank problems in order of importance and then analytical techniques such as problem statement triggers, Duncker diagrams, KTDA, and KTPPA, we were able to prioritize and propose the most feasible solutions to our initial problems.

Problem Number	Priority	Problem	Feasible Solution
1	1	Co-workers do not take responsibility for job tasks	Task checklists and rewards system
2	2	Bottlenecks occur because not enough workers are present during each shift	Purchase more equipment
3	3	Employees do not receive enough training for tasks	Hire competent people and provide more efficient training

Implementing the Solution

We recommended that the manager of Bakery Café implement these solutions to the identified problems. The situation appraisal showed that the problem of co-workers not taking responsibility for job tasks was the most important and needed immediate action. We proposed setting up task checklists at appropriate stations detailing the task itself as well as any prep or clean-up duties associated with it. To implement this solution, we would need to create and laminate checklists detailing various tasks. If certain tasks consistently are not being done or are being left for later shifts, then a sheet could also be created that employees would have to initial when the tasks are completed. This solution has a very low cost and is easy to implement.

Another idea was to designate one of the workers on each shift as the "checkout" employee. Before other workers end their shifts, they would have to check out with this employee, who would make sure that all clean-up and prep tasks were complete before the next shift began. This person would preferably be one of the more responsible employees during the shift but should also change from day to day so that multiple associates are exposed to this responsibility. This solution would increase the general knowledge of the tasks that need to be done because the "checkout" employee would have to know all of the tasks for several different areas.

We decided that the problem of not enough workers could be solved by purchasing more equipment. It seemed that a bottleneck occurred during busy hours of the day owing to the high demand placed on the espresso machine. This problem seemed to result more from the lack of equipment than from a lack of workers. Purchasing a second espresso machine would help shorten customer waiting times and increase customers' satisfaction. The purchase of a second espresso machine might not be economically feasible, however. The cost of another machine might be offset by increased customer and worker satisfaction, but it probably would not increase revenues much. Also, the store might not have space for a second machine or adding a machine might mean sacrificing another device.

An alternative solution would be to create a sign that displays the current drink order number. This sign would give customers a better idea of what to expect in terms of a waiting time and reduce the stress of impatient customers. This expectation would allow them to be more relaxed and free to enjoy the Bakery Café's atmosphere—an area that Bakery Café excels in.

The problem of employees not receiving enough training for tasks is a bit more complex because most training programs are formulated at the corporate level. If we assume that the program itself is not flawed, then either training needs to be implemented more efficiently or it needs to go on for a longer time. Perhaps associates are not being cross-trained thoroughly enough to perform efficiently, or perhaps associates are being trained at too broad a level and are not going deep enough in certain areas. Associates may also be receiving adequate training, but then specialize in certain areas based on which tasks they are assigned. To solve this problem, employees could be given refresher courses in areas where they do not work often.

Evaluation

We would need to follow up with surveys or (preferably) short interviews to see how employees react to the changes made. The changes made should increase worker and customer satisfaction. Evaluation would be a task for the managers at Bakery Café because they see how things run on a day-to-day basis.

SUMMARY

Solving problems can be fun and add spice to an otherwise bland existence. The techniques discussed in this book for defining and solving real-world problems are tried and true, and they can make you a better problem solver. Use the materials in this book to lessen your problem-solving anxiety.

- Take a proactive approach not only to problem solving, but also to your life.
- Have a vision of what you want to accomplish in your organization and in your life, and make sure all of the steps you take head in the right direction.
- Don't be afraid to fail. Most of the time we can learn more from our failures than we do from our successes.
- Observe good problem solvers. Ask them questions and brainstorm with them.
- Don't dismiss the ideas of others out of hand, but instead leverage them—they may trigger the correct solution.
- Get in the habit of planning your time and prioritizing the tasks you must accomplish.
- Practice using the tools and techniques presented in the book. A tool becomes useful only when you are comfortable and familiar with it. Some techniques may sound silly or cumbersome. Don't worry: The same technique will not work for *everyone* and *every* situation. Choose what works best for you.
- Don't force-fit a certain technique that you have used before. If it's not working, choose another technique or move on.
- When you have finished a task, reflect on it. Evaluate it. Is it the best you could have done? If not, can you still improve upon it?

Every time you practice the problem-solving techniques we have presented, you will become an even better problem solver.

WEB-SITE MATERIAL (WWW.UMICH.EDU/~SCPS)

- **Learning Resources**
  Summary Notes

INSTRUCTOR'S SOLUTION MANUAL

Because of electronic copyright restrictions, the following material could not go on the Web site but is included in the *Instructor's Solution Manual*:

- Sears Catalog
- Coors Beer

These case studies, which appeared in the first edition of this textbook, can be found in the *Instructor's Solution Manual* along with two case studies from the term projects assigned in the University of Michigan's senior-level Engineering 405 course, "Problem Solving, Troubleshooting, and Making the Transition to the Workplace":

- Alarm Clock Design
- Problems in the Dorms

REFERENCES

1. Fink, Arlene, *How to Conduct Surveys: A Step by Step Guide,* 3rd ed., Sage Publications, Thousand Oaks, CA, 2005.
2. Oishi, Sabine Mertens, *How to Conduct In-Person Interviews for Surveys,* 2nd ed., Sage Publications, Thousand Oaks, CA, 2002.
3. Creswell, John W., *Research Design: Qualitative and Quantitative Approaches,* Sage Publications, Thousand Oaks, CA, 1994.
4. Frary, Robert B., "Hints for Designing Effective Questionnaires," *Practical Assessment, Research, and Evaluation,* 5, 3, 1996. Retrieved October 19, 2006, from http://pareonline.net/getvn.asp?v=5&n=3.
5. Hill, Nick, "Tips for Developing an Effective Questionnaire," 2004, www.streetdirectory.com/travel_guide/2244/computers_and_the_internet/tips_for_developing_an_effective_questionnaire.html.
6. "Survey and Questionnaire Design," www.statpac.com/surveys/index.htm.
7. "Survey and Questionnaire Construction," www.tele.sunyit.edu/TEL598sur.html.

EXERCISES

Choose your group term project from the list below.

Business Projects

12.1. Select a business of your own choosing (e.g., your local lumberyard, a bakery, a photo shop, or a hardware store) and get approval from the manager/owner to determine the problems faced by the employees. Try to discover the "unspoken" problems as well as those problems you may have uncovered in focus groups or from a survey of the restaurant staff and customers. For each and every problem you uncover, generate a number of solutions, pick the best solution for each problem and then prepare an implementation plan. Groups of four students from the Fall 2013 class of Engineering 405 used Panera Bread, Espresso Royale, Zoup, Bear Claw Coffee, and the University of Michigan Dorms as their clients for their term project. Previous classes included clients Qdoba, Ace Hardware, Gross Lighting, and Better Brigade. At the end of the term each group gave a 40-minute presentation to the class and the clients. The clients' evaluation counted for 40–50% of the grade for the course. This exercise can be done in groups or individually.

12.2. Review the reference material and Web sites on Survey Design given in references 1 through 7 on page 330. Prepare a survey to interview 50 students who live in one of the dormitories at your school to determine the problems they have living in the dorm. Try to discover the "unspoken" problems as well as those problems you may have uncovered in interviews. For each and every problem you uncover, generate a number of solutions, pick the best solution for each problem, and then prepare an implementation plan. See the Engineering 405 syllabus on the Web site.

12.3. Choose a popular consumer product (e.g., hand soap dispenser, either foam or liquid) currently on the market that you believe could be improved. Review the reference material and Web sites on Survey Design given in references 1 through 7 on page 330. Prepare a survey to interview 25 people, asking them what they don't like about the product (e.g., too much soap dispensed), and then generate and evaluate a number of solutions to each problem.

Other Projects

12.4. Research a technological need in Rwanda, Africa (e.g., turning solid animal and human waste into fuel). Define the problem, collect information, and outline the solution. Include a Gantt chart and a potential problem analysis as part of your analysis. Prepare a proposal to go to Rwanda consistent with that of Engineers Without Borders (EWB). Use the Web to obtain information about EWB. You can find additional information about investment opportunities in Rwanda on the Web at www.thvf.com/.

12.5. Determine the major problems that would affect students, dorms, and businesses should the bird flu pandemic reach your hometown. Which businesses would suffer the most if the bird flu reaches the United States?

12.6. Research one of the following topics.

A. Global warming

B. Sustainability

C. Water shortage/conservation

Prepare and carry out a campaign to make people aware of your topic. Develop a Web site that gives the scientific facts, concerns, and consequences of your topic. If nothing is done to address the issue, what will be the consequences?

INDEX

PRENTICE HALL

REGISTER

THIS PRODUCT

informit.com/register

Register the Addison-Wesley, Exam Cram, Prentice Hall, Que, and Sams products you own to unlock great benefits.

To begin the registration process, simply go to **informit.com/register** to sign in or create an account. You will then be prompted to enter the 10- or 13-digit ISBN that appears on the back cover of your product.

Registering your products can unlock the following benefits:

- Access to supplemental content, including bonus chapters, source code, or project files.
- A coupon to be used on your next purchase.

Registration benefits vary by product. Benefits will be listed on your Account page under Registered Products.

About InformIT — THE TRUSTED TECHNOLOGY LEARNING SOURCE

INFORMIT IS HOME TO THE LEADING TECHNOLOGY PUBLISHING IMPRINTS Addison-Wesley Professional, Cisco Press, Exam Cram, IBM Press, Prentice Hall Professional, Que, and Sams. Here you will gain access to quality and trusted content and resources from the authors, creators, innovators, and leaders of technology. Whether you're looking for a book on a new technology, a helpful article, timely newsletters, or access to the Safari Books Online digital library, InformIT has a solution for you.

informIT.com

THE TRUSTED TECHNOLOGY LEARNING SOURCE

Addison-Wesley | Cisco Press | Exam Cram
IBM Press | Que | Prentice Hall | Sams

SAFARI BOOKS ONLINE

informIT.com
THE TRUSTED TECHNOLOGY LEARNING SOURCE

InformIT is a brand of Pearson and the online presence for the world's leading technology publishers. It's your source for reliable and qualified content and knowledge, providing access to the top brands, authors, and contributors from the tech community.

Addison-Wesley   Cisco Press   EXAM/CRAM   IBM Press.   QUe   PRENTICE HALL   SAMS   Safari Books Online

LearnIT at InformIT

Looking for a book, eBook, or training video on a new technology? Seeking timely and relevant information and tutorials? Looking for expert opinions, advice, and tips? **InformIT has the solution.**

- Learn about new releases and special promotions by subscribing to a wide variety of newsletters. Visit **informit.com/newsletters**.

- Access FREE podcasts from experts at **informit.com/podcasts**.

- Read the latest author articles and sample chapters at **informit.com/articles**.

- Access thousands of books and videos in the Safari Books Online digital library at **safari.informit.com**.

- Get tips from expert blogs at **informit.com/blogs**.

Visit **informit.com/learn** to discover all the ways you can access the hottest technology content.

Are You Part of the IT Crowd?

Connect with Pearson authors and editors via RSS feeds, Facebook, Twitter, YouTube, and more! Visit **informit.com/socialconnect**.

informIT.com
THE TRUSTED TECHNOLOGY LEARNING SOURCE   PEARSON

Addison-Wesley   Cisco Press   EXAM/CRAM   IBM Press.   QUe   PRENTICE HALL   SAMS   Safari Books Online